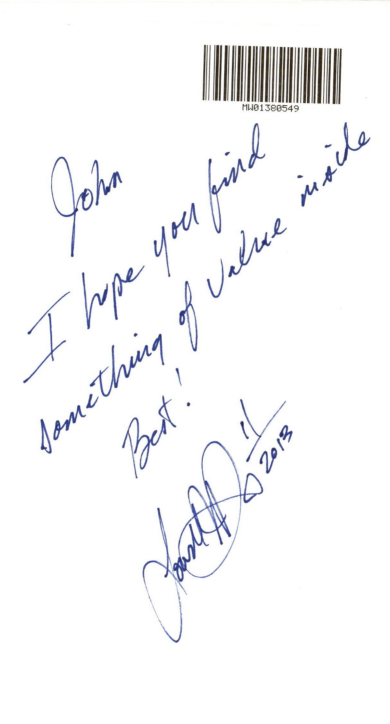

Keep It Simple and Sustainable

Lean Leadership Methods that Build Sustainment

Lowell J. Puls

authorHOUSE®

AuthorHouse™
1663 Liberty Drive
Bloomington, IN 47403
www.authorhouse.com
Phone: 1-800-839-8640

© 2013 Lowell J. Puls. All rights reserved.

No part of this book may be reproduced, stored in a retrieval system, or
transmitted by any means without the written permission of the author.

Published by AuthorHouse 8/26/2013

ISBN: 978-1-4817-7109-2 (sc)
ISBN: 978-1-4817-7110-8 (hc)
ISBN: 978-1-4817-7111-5 (e)

Library of Congress Control Number: 2013911688

Any people depicted in stock imagery provided by Thinkstock are models,
and such images are being used for illustrative purposes only.
Certain stock imagery © Thinkstock.

This book is printed on acid-free paper.

Because of the dynamic nature of the Internet, any web addresses or links contained in
this book may have changed since publication and may no longer be valid. The views
expressed in this work are solely those of the author and do not necessarily reflect the views
of the publisher, and the publisher hereby disclaims any responsibility for them.

Foreword
"The truest definition of waste is working to become more efficient at something you shouldn't be doing in the first place!"

My father often uses the phrase "Keep It Simple, Stupid" and it's one I've used occasionally as well. While looking for a name for this book on sustainment, a modified version of the phrase popped into my head, and before a word had ever been written, the book had both a name and a theme.

A close friend of mine once said that leaders are smart, complex people who don't believe simplification is possible. Although he's partially right, most of my experience suggests it isn't so much that they don't believe simplification is possible as it is that they aren't very good at simplifying what they're trying to accomplish. The difficulty seems to come from their approach at responding to and solving problems in their business. Many leaders attempt to monitor and control excessive levels of detail, leaving them overtasked and impeding assistance from more appropriate subordinates who should own the task. They develop "solution lock," or a tendency to gravitate to a single solution for a problem earlier than the facts would suggest is appropriate. Even worse, they cling to that solution even when it begins to prove sub-optimum.

One of the best examples of simplification comes from an old training room joke many have heard about a semi-truck that gets wedged under an overpass. While the police, fire, and civil personnel are at a stalemate as to how get the truck out, a ten year old boy walks up and asks, "Why don't you just let some air out of the tires?"

There is yet another story that similarly illustrates this point. A great bow hunter heard of a teenager with the reputation as the greatest archer in the state, someone who never missed the bull's eye. So the famous hunter went to visit this boy and asked to see his practice range. The boy took him into the family back yard, encircled by a wooden fence with more than a dozen targets on it. Smack dab in the center of each target was an arrow. The man was amazed. "You always get the arrow right in the center of the bull's eye?" he asked. The boy nodded in assurance. When asked to demonstrate his skill, the boy fetched his bow and inserted an arrow, drew back and sent the arrow into the wooden fence. He then ran inside his garage, returning with a bucket of paint, and painted a bull's eye around the arrow. When done, he turned to the man and announced, "It works every time."

The objective of this book is to identify management behaviors that complicate initiatives and eliminate them and in the process, free up improvements in performance and sustain them. It's been written in short form to preserve the purity of its true message – that simplification is necessary in every situation, and finding it will make your solutions both more effective and transferable. Originally targeted ten pages per chapter including a "story page," it constantly flirted with being much longer. True to form, I continued to go back and simplify it. Although there are a few concepts that might not seem to be explained thoroughly enough, it's reflective of a conscious intent to provide the simplicity of the approach without prescriptive detail, leaving you the latitude of filling in the specifics of your own business's needs.

It can be basic considered human nature to overcomplicate every element of running a business, from the basic concepts of strategizing; planning, execution, and adjustment, to accounting, engineering, customer service, and human resources. These individual business functions can assume a life, goal, and agenda all their own, usually connected to the business objectives but almost always disconnected in some way from the essential mission of being the very best at servicing the needs of a target market. Applying the concepts in a basic step by step fashion will enable you to find your own approach and adapt to each business condition without limitations. In any successful enterprise, the process solutions can be flexible but the process disciplines can't be compromised. Following the simplicity of Lean Enterprises' "standard work," the chapters are written in a "major step" (the key concept being communicated), "key points" (what is required to assure a perfect quality result), and the "reason why" (because communicating or explaining the intent is critical to selling acceptance of the disciplines). If this book doesn't help simplify your approach to aligning activities and improving performance in your own business, call me for a translation.

Thirty-eight years in manufacturing and business have led me to believe that although I haven't *seen* everything, it seems certain I've *stood next to it*. My experiences have amassed many stories that are fun to tell, and this book is woven from a number of true stories to provide supporting examples. A few of those who shared them with me might even recognize the situations.

More than anything, after working to transform a variety of companies, it seemed timely to make a statement about how to sustain improvement. First, you must connect clear targets for improvement to the business mission, not to the careers of those in leadership roles. Second, if your leadership style isn't respectful of the human element for its power to propel your cause, the improvements will under-achieve, causing the organization to sustain them minimally, or possibly even lose its competitive edge. Third, a leader must take great care to ensure that his objectives are pure and that the successes are shared by everyone involved. Management integrity is foundational to sustainment, and high turnover at the leadership level will always undermine continuity. If your leadership development program uses a ticket punch approach to diversify experience, you haven't even scratched the surface of its real cost impact to your business.

I would like to acknowledge those who have worked with me and directly or indirectly taught me much of what you will read about! Their willingness to share wisdom and experience was always the greatest gift. We've worked together to transform companies, save and rebuild customer relationships, and blaze new trails, all the while time working for the betterment of the team. Among the many who have taught me the most important lessons are: Rod Garfield, Gene Holman, Ken Davis, Jim Sweetnam, Sandy Cutler, Craig Henderson, Barbara Drawbridge, Caryn Myrice, Evan Arrowsmith, Ted Hall, and Dan Griesbach, among many others. Finally, my family also taught me, as I sacrificed too much of my time with them to work, that never ending love really exists. Each of them continues to teach me today, an exchange that will hopefully continue for a very long time.

Enjoy the book! After that, go paint some bull's eyes on your objectives and let some air out of your tires to get moving!

Table of Contents

CHAPTER 1: Lead With Vision 1
Walking the fine line between vision and hallucination to build a Winning Strategy

CHAPTER 2: Leadership Response-Ability 17
Setting the tone for a more Responsive Organization

CHAPTER 3: Keeping things RADically Simple 35
Using the Success Equation; Results = Approach + Deployment

CHAPTER 4: Organizing for Change 47
Align Your Structure with High Performance Staffing

**CHAPTER 5: Things That Will Kill You, Things That Can Eat
You, Things That Make You Happy!** 61
Realistic priority management

CHAPTER 6: Creating "Behavioral Pull" 73
People Systems that Trigger Team Achievement

CHAPTER 7: Takt-ical Management 87
Using the Lean Tools to Build Improvement

CHAPTER 8: The ABC's of SQCD+T 107
Simplified Metrics that Self-Improve

CHAPTER 9: "En-gage!"...121
Involve to Engage and Create Sustaining Behaviors

CHAPTER 10: Consistent-ize 133
Business Systems that Secure Long-Term Performance

CHAPTER 11: Connecting the Dots 145
Adding it all up for Extreme Business Excellence

Chapter 1:
Lead With Vision
Walking the fine line between vision and hallucination to build a Winning Strategy

1) Situation Analysis
2) Defining Your Vision
3) Communicate the Vision

"Lowell, when you get done with *this* job, you're going to have that big journeyman's tool box!" Jon, the chief operating officer, was referring to the skills I would develop through the course of the assignment. Something told me he wasn't exaggerating. This job was a true test for anyone at any level of experience. Having spent the past two years working directly for him, flying all over the world and solving business-critical issues, such as correcting matters of meeting market supply needs, cleaning up quality spills, and preparing and executing transformational strategic plans, I certainly felt ready – hopeful those experiences were preparatory enough for the challenge.

He was handing me the keys to a troubled $300M business, besieged with both customer and labor challenges, while losing business quickly as a consequence of poor customer response. Somewhere along the way, the associates had lost their connection to the true mission of delivering high quality products within the customers' expected timeframe. Both its processes and people would need some serious "Lean Thinking" to get things moving forward again, and to recognize on a daily basis, that their business mission was an extension of satisfying customer needs. Currently, they were turned inward, tracking metrics that were more effective at rationalizing their poor performance rather than striving to meet the expectations being demanded by the customers. Even their strategies regarding production and sales seemed disconnected from the market itself.

It had been a whirlwind courtship. The previous president had abruptly retired after only a year in the role, leaving the leadership position vacant. A mere four days passed from our initial discussion to my acceptance of the job.

The very next Monday, I found myself sitting alone in the division headquarters' waiting room. The management team was assembled in an adjacent conference room and being briefed on the change. As the door opened, Jon waved to me: "Welcome to the team," he said loudly, as I walked in. After a short round of introductions, he continued. "These folks understand the challenges and are ready to do whatever it takes to turn this business into a top performer." After a few moments of small talk he exited the room and left me to address the staff. Already armed with Jon's opinions on the business condition, the contributing factors, and the caliber of many of the team members, it was time to form opinions of my own.

"My challenge for the coming weeks is to have an immediate impact on customer service levels, while at the same time circling us back to budget. I'd like to get to know all of you, and plan to meet privately with each of you within the next few days to get your views on the causes of these severe performance problems. From there, I'll summarize what I hear, rephrase the details to preserve confidentiality, and feed a summary back to you as a team. We'll then try to determine our best course of corrective action. It's critically important that you are open and honest about the problems, or else we'll waste time pursuing the wrong solutions. Until then, are there any questions?"

"Do you anticipate making any immediate changes?" asked Mary, the Director of Human Resources.

"Not at this time," I responded. "It's essential for the entire organization to develop confidence in this team and in me, because we'll really need everyone's trust going forward. That means we can't afford to flail around and make mistakes. Once the interview process is complete, we'll align on a plan and execute it with the appropriate urgency. That said, however, some changes are inevitable. Are there other questions?"

Since there were no takers for the second offer, we scheduled the individual interview rounds before adjourning.

Organizations that get into trouble usually do so for myriad reasons, but the results are typically anything but varied – they're simply bad. Like others I'd seen, this business had entered "the death spiral," a place where performance had deteriorated to a point that everyone was so immersed in problem resolution that essential business functions were being neglected, causing further performance deterioration.

I knew there would be some suspicion and anxiety regarding my private interviews. Ronald Reagan once said the most feared words any organization could hear were, "I'm from the government, and I'm here to help you." Well, announcing, "I'm a complete outsider, and I'm here to turn your company around" must run a close second. Nevertheless, I conducted the interviews, listened carefully, took some notes, asked a series of scripted questions, and then thanked each person. As might be expected, some were cautious and tight-lipped. Others were bitter and quick to lash out over grievances real or imagined. Still others were transparent, genuine, insightful, and honest.

A week later we met for our first staff meeting to review the interview results. I sensed the team seemed a bit nervous, not knowing how their input would be used or how candid I might be regarding what was said. They were in for a surprise.

"Welcome, everyone!" I opened positively and with a smile to try to ease things a bit. "It's essential that you understand how simple my motives are. First, let's dismiss any belief that everything around you is completely broken. This is an important message to spread to your teams as well, because as we dig in to make changes, we'll need everyone to believe in what we're going to be doing. Without this, the employees will view our efforts as hopeless, and they'll quickly return to 'muscling' processes to maintain the minimal performance they're already getting, thus guaranteeing defeat. Every organization has things they do well, and we'll have to identify and preserve those, both to save time by not fixing things that don't need to be fixed, but also to sustain a level of pride in doing something well. Next we'll reach consensus on the leading issues that will require us to focus our improvement efforts. Once we've accomplished that, we can establish an agenda to attack the initial target areas."

Nearly everyone had indicated that we weren't servicing our customers adequately nor fulfilling the commitments *we* were making to them. Our market research also confirmed the fact that we weren't delivering what they wanted or expected. As I engaged the team in a standard "5 WHY" approach to our market issues, the input which the various team members had confidentially given me was selectively interjected. This helped us narrow in on root causes so that we could proceed toward addressing the primary issues.

"Why are customer service levels so poor?" I asked.

"Because their demand is so erratic that we can't forecast it properly," Geoff, the operations director fired back.

"Well, yeah, but even if we forecast it correctly," injected Janet, our Director of Materials, "our manufacturing time is longer than the market lead time allows anyway."

I asked, "What *is* the market lead time?"

"Usually about sixteen weeks," said Dick, VP of Sales.

"Says who? How do we know?"

Dick looked somewhat confused. "Uh, because that's the lead time our competitors quote. It's kind of always been that way."

"Really? Okay, back up a minute." I paused. "Do we *always* quote 16 weeks?"

"Yes, we do," Dick stated flatly, as though this was common knowledge.

"I'm aware that a number of our products are custom built. Can anyone tell me what percentage?"

Dick shrugged. "Percentage? Shoot, everything we do is custom."

I looked over at John, our VP of Engineering. "Your people touch *every* order? Gees, what's your typical process time?"

"It sometimes can be done in a week, but other times it takes a month. It all depends on the level of engineering required."

I weighed that a moment. "If every order is custom to some degree, then are they just custom-assembled, or do they require specially designed parts as well?"

"Special parts are often required and usually have two-to-four weeks of procurement time," Janet answered.

I turned to her. "What are the longest lead time parts we buy for these products?"

"It's the Asian stuff, so count on six months for large castings and around18 months on proprietary assemblies."

KISS: Keep It Simple and Sustainable

I tapped a few notes to myself on my tablet computer. "Okay, Geoff, can you give me some kind of idea what your manufacturing time is like?"

"What, for normal runs? Sure. Usually, we can complete Product A in four weeks if it has minimal customization. Product C is always customized and can take up to 26 weeks to make and ship."

I sat quietly for a moment, tapping more notes and organizing my thoughts. Finally, I looked up at the whole team. "So, if we take one to four weeks to engineer an order, then two weeks to 18 months to buy the stuff that goes in our products, and finally four to 26 weeks to produce it, how on earth do we deliver in the 16 weeks quoted?"

"We don't," Dick answered, somewhat sheepishly. "Leastwise, not very often."

"*How often?*" I asked him pointedly.

He swallowed. "Well...uh...other than Product A, I can't remember ever."

I looked from one face to the next around the table. "Are you serious? Then why on earth do we quote it that way?"

"Have to," said Dick defensively. "If we didn't, we wouldn't get the order!"

I just stared at him, open-mouthed. This comedy of errors was beginning to circle back on itself. I was beginning to feel as though I was doing the Abbot and Costello "Who's on first?" routine. The problem was, this was a manufacturing business, not a stand-up comedy club. Well...at least that was what it was *supposed* to be.

I forced myself to suppress my angst and stay calm. "Okay, let's go back to *why*. Why are our customer service levels so poor? How did this happen?"

"Our systems are broken!" Mary from Human Resources said, suddenly entering the fray. Her flushed cheeks matched her curly red hair. She had been reticent to talk to me at first during our private appointment, but once I'd gained her confidence, she'd rattled off a checklist of things that had been sticking in her craw for a long time. She wasn't a rabble rouser. She had given 14 years to the company and was fiercely loyal to it. Seeing it flounder was upsetting to her personally and professionally.

"Thanks for that gut honesty, Mary. I think your confession of reality just opened a door that may lead to some potential answers for us. Okay, I want the same from the rest of you. We're behind closed doors, and no one has a tape recorder on. Let's hear it. Lay it out. Spill your guts."

Walking through the remaining "whys" took a couple of hours, and I could see the frustration and concern building in each of them. However, at the same time, we successfully identified our top priority, to regain our former high-quality of customer service. Ultimately, it became obvious to everyone in that room that if our company couldn't make good on its commitments to delivery schedules, production excellence, and follow-up service, nothing else mattered. Forget the fancy advertising slogans, lucrative contracts, crisp uniforms, shiny delivery trucks, and service to the community. Nope, without satisfied customers, none of the rest of it made a bit of difference. On that, we agreed, and having formed that consensus, I gave them a two hour break to grab some lunch and to check in at their offices. At the same time they were tasked to mull over what we'd covered that morning and to come back in the afternoon with additional comments, suggestions, or even rebuttals.

* * *

When we re-convened, they were much more focused. Several of them had returned with tablet devices upon which they had typed notes. I was anticipating some lively interchanges and was not disappointed.

"Now, let's try to define the top three or four reasons why we're not meeting the needs of our clientele," I said.

"We can't forecast the business!" Jolene, the financial director, blurted out. She had been somewhat subdued during the morning session, but I now saw that having had some time to mentally process the discussion approach, Jolene was ready to make her voice heard.

"Why?" I prodded. "And I'm not just asking Jolene."

"Because Sales never commits to anything until it's a signed order – then it suddenly becomes a five alarm fire," Geoff half mumbled, not disguising his disdain.

Dick winced at that. "Look, my team isn't made up of mind readers," he retorted sharply. "We can't tell when the customer is going to place an order, much less how large it's going to be. We're the ones who keep the cash flow coming into this place. It's your job to see the orders get filled."

KISS: Keep It Simple and Sustainable

I raised my hands for peace. "All right, all right, let's look at that a minute, folks. At the present time, how do you plan and purchase those materials that are out six to 18 months?"

Janet intervened. "There's no immediacy to anything. We look at what the trends have been for the past two or three years, and we hope that we'll be similarly fortunate for this year. So, we place orders based on history, not on actual orders-in-hand. The results are chaos. We're either way too short on one product, or we have a two-year stockpile of another. It's like rolling dice and it drives me crazy."

Geoff gave a sarcastic chuckle. "Drives *you* crazy? Man, on the factory floor it's feast or famine. We either have big orders and no materials, or else we have warehouses full of materials that no one wants any of. My people will go from working double shifts and running up the overtime, to being sent home for nothing to do."

"So, what efforts have you made to try to level your production so that the factory and the materials teams can work on a level basis?" My prodding was working.

"We build Products A and D to inventory in the fall and winter," Janet and Dick said almost simultaneously.

"How can you do that if nearly everything is custom?" They could sense my confusion.

"We build base units and then rework them as needed to make them order-specific," Geoff said.

"How do you cover the extra cost?"

"We don't! And sometimes the rework is extensive, hitting our margins. " John confessed.

"Okay, is it safe to say that much of this could be eliminated if your S&OP went through a major evolution to stabilize your planning process?" They looked at me quizzically. "It's short for Sales and Operations Planning."

"But we don't currently have a process by that name," Janet stated.

"Sure you do!" I answered back. "Only it's called Sold - and Operations Panics!" Geoff chuckled as I said it

As I'd expected, the list quickly expanded beyond the number one problem of forecasting. Soon they were also listing execution and engineer-ing-derived product launch issues. Additionally, we uncovered a number of more minor procedural items, such as multi-tiered approval timing and

7

system functionality issues, but we were able to agree quickly on ways to eliminate them, neutralize their impact, or cast them aside in order to our focus on the larger problems. By this point we had successfully outlined our top organizational priority as "Delivering to our customers in *their* required timeframe," which blossomed into three initial projects.

Using an approach I call 3C's (**C**ontain the problem / **C**orrect the Problem / **C**ontinuously Improve beyond it) to address our performance, we began by sorting through the steps required to contain our operational issues and <u>stop</u> them from impacting the customer, regardless of how it impacted us internally. The team worked diligently, but it took some time and effort to identify containment on the thorniest problems, as well as considerable patience in getting the team to drill down to root cause. As the team improved at building a complete, focused action plan (addressing what we knew and then highlighting what we needed to know more of) for each issue, before moving to the next problem, we began to pick up momentum. The group was just beginning to visualize a path toward resolving our issues for the customer.

Our second organizational imperative turned out to be returning to budget, and here the discussion proceeded more quickly. Most of the causes of budget noncompliance were driven by corrective or containment actions resulting from our customer satisfaction problems. The project priorities came readily, and we quickly established a path to improvement.

We continued this process through the course of each of the three improvement initiatives. Two of them addressed our customer performance, but one was financial. From there, we tentatively assigned the identified corrective action steps across all of the functions, then agreed on a plan to communicate the new short-term initiatives and their expected impact to the broader organization.

"Now, let's talk about how we're going to pass along this information to all of the folks who report to each of you. I would appreciate some suggestions."

"They will see some of what we're proposing as having been tried before," Mary warned. "To avoid having them tune out, we need to come up with some fresh expressions or examples or ways of explaining our goals. Same-ol, same-ol isn't going to hack it."

KISS: Keep It Simple and Sustainable

"I'm in your camp on that," I concurred. "But have a suggestion on how we can present some potentially redundant material in a new way."

"And that would be....?" asked Mary.

"By doing instead of talking," I said. "You know the old adage about actions speaking louder than words. In realigning a business, that's certainly true. So, yes, we've got to sell our coworkers on the idea that we've carefully thought through these changes, and have to convince them that we are under pressure regarding the time in which a turnaround must generate success. Lectures and pep talks and warnings aren't going to be effective. We need to walk the talk."

Mary grinned and said, "You want us to convey to them that our message is that they should do as we do."

I nodded. "That's the ticket. Lead by example. So, right now, I want each of you to write out one resolution regarding a behavior of yours that you will commit to changing as a clear example of how everyone is going to have to put the company's needs first."

"Like what?" asked Dick.

"Look down at the notes you've been taking all day," I said. "Focus on something significant. Off the top of my head, I'd say a good resolution for you would be to avow to discuss any major order with the rest of the team before you commit to a due date."

Dick squinted. "We've tried that before, and I've never gotten a decent response in time to give the customer an answer. What's gonna change that?"

"New thinking and team commitment!" I insisted. "Okay, so from today on, the rest of the team and I will commit to a 24-hour response to any customer request. Will that work?"

Although Geoff and Janet squirmed a bit, Dick nodded. At least in theory, we were all on the same page.

Eyes *WIDE* Open for a Change

Even great organizations can fail to address major problems in their business because of an inability to identify their causes and then focus on corrective measures. I think sometimes it's because they fear a major improvement effort might compromise an already tenuous status quo. It's almost as if they are always playing defense, trying not to lose rather than to make an effort to win. They will diligently attack symptoms as if they're independent events rather than the consequences of deeper problems, causing them to employ remedies that can create even more troubles. An example seen frequently is one of adding an approval step for the procurement of materials as a measure to reduce inventory, when the real problem is either broken forecasting or glitches in the ordering process. Rather than correcting the process problem, a containment style oversight step is imposed, which only dulls the response of the overall business. It actually compounds the problem rather than correcting it and, more often than not forces people to work around the system, breaking things down even further. I call this the "death spiral."

1.1 Situation Assessment

There is a fine line between vision and hallucination, and the only difference is a plan. I say this to stress that no matter how far-fetched a vision might appear, if it can be rooted to reality through careful and diligent planning, then it can be elevated from a seeming fantasy into something achievable. Translating a vision into an executable plan allows leaders to walk on the correct side of that fine line and plays a factor in whether or not people will follow them.

Reconnoitering – Is an old pathfinders' term that refers to the practice of using multiple benchmarks within one's surroundings to approximate a current location and to determine the correct direction to head in. In business, this requires stepping back from the action taking place, in order to gain a broader view of your entire surroundings. By "walking" the business and looking at the visual indicators that can be observed -- both good and bad – you can develop a rough list of things that need to change and those that should be left alone. Remember though that these are symptoms only, and root causes will be harder to discover. Your list can be reviewed regularly and modified as often as necessary in order to stay on a course of improvement.

Triangulation - While it's senior leadership's responsibility to establish a vision for the business, rooting it to reality requires a thorough situation assessment. Triangulation among the current performance of the business, your competitors' performance, and the known market benchmarks for customer expectations, will enable an assessment that is more securely based in fact. Positioning a business amidst these three points can establish the standard for performance excellence that applies to it and helps identify any gaps that need closure. Once it's defined, navigating to a vision that challenges the organization to pursue what's possible, while at the same time stretching it toward a set of breakthrough objectives, becomes a highly credible task supported by three legs of reality.

Alignment. An organization will inherently sense the difference between vision and hallucination. When the goals, the execution plans, and the deployment all fail to line up, it causes a lack of believability that muddies the clarity of your vision, and must be corrected before commitment and buy-in can be achieved. Oftentimes, even when the details are well defined, the attainability of stretched objectives fails to be clear to the associates without some additional definition and salesmanship. Just like a tree, the loftiness of a company's growth is controlled by the depths of its roots. The ability to reach high objectives is limited by how well grounded the logic and supporting information can sustain the vision! This information not only forms the basis for the newly established goals, but it also serves as an outline for communicating its purpose.

360° View. However initiatives get started, it's essential that the vision be formed from a blend of external drivers (markets and competitors) and internal needs (closing performance gaps). In history, a good example of a lack of vision is the construction of the Maginot Line, a series of fixed battlements France built on its border with Germany after World War I. The French intended the wall to keep the Germans from overrunning them in any future dispute as they had in WWI, but the strategy was flawed because it focused on war tactics in their current state, failing to consider how they might evolve with future changes in technology and tactics. By the time of its completion, new technologies (tanks and planes) and the related battle tactics had already evolved in a way that effectively made fortressed walls obsolete. When World War II broke out, the Germans made an easy task of flying over or blasting through the defenses and conquering France. The lesson: a correct vision not only attends to the current state but is adjusted actively to include the future as well, in the process anticipating challenges and opportunities, or even using technologies that may not exist in the present.

1.2 Defining Your Vision: Visualize a path you can follow!

A vision can originate from something as simple as a high level philosophical statement, or it can be built through a team-based tactical exercise. For example, stating something like, "I want us to be the market leader" is similar, but different from, "We need to move ourselves from the middle tier to the top tier of our competitive peer group." Though both can mean relatively the same thing, the latter suggests the existence of more substance than the first. No matter how it's formulated, the vision statement must offer a sufficient amount of information to imply that the vision makes sense. A business may have a vision to become the market leader, or to move its perceived identity from that of a "parts maker" upmarket into a "systems integrator," but any vision's objective should be realistic enough to appear attainable. At times this may necessitate breaking the macro-vision into smaller micro-steps. Creating a rational and functional vision requires the assimilation of data, substantial preparatory work, and a logical execution plan.

Untouchables. Framing the vision begins by having the leadership team define those principles of the organization that cannot be altered. These will include elements that make up the current state, such as product quality, a tradition of customer service, market or brand value elements, business ethics and personal morality, and possibly even the brand logo or mission statement. Why are they untouchable? If they are significant enough to form the retained value of the old business, they should also be foundational in the transformation. Further, their continuity serves as a stable vantage point from which people can stay connected to those things that factored into their original success. The reality is that there will be enough to change, so it's unwise to alter things that aren't broken, saving time and resources for the true needs of the business.

Must Changes. With the untouchables defined, the list of critical changes becomes clearer and easier to identify: improvement of competitive position or core competencies, new product development and launch, and possibly even order management. All of them are capable of transforming business performance, market position, or brand integrity in line with the achievement of the visionary objectives. Once each factor is evaluated for priority and potential impact, they, together, will comprise the must-change list that supports the overarching strategy. As the must-change list is drafted, recheck it against the untouchables to make sure that none of them pose a barrier for progress. Such conflicts should be resolved delicately in order to preserve the most important elements of each.

Integration. Translating your vision into an executable strategy is a far more tactical step that should enlist the entire organization to help build buy-in, enhance organizational skills, and ensure that all valid input is considered. Objectives for growth, profitability, or market share should be based on market information, customer feedback, business performance, benchmarking information, and even the organizations culture. Having the team "mine" this detail helps ground the vision in their eyes, preventing it from degrading into hallucination. The facts and data used to connect the vision to reality also formulate a basis for communication during deployment to the organization.

Clarity. To better understand the need to provide clarity of purpose, consider one company whose vision came from the top in the form of a blanket "percentage of improvement task" over a strategic planning period of three years. The business unit teams set improvement targets based upon the prior year's performance and in line with the corporate objectives – both the metric targets as well as the program goals. Because the corporation's annual plans were always structured as the first stepping stone leading toward achievement of the 3-year strategic plan (and adjusted annually based on actual attainment as well as for market changes), the resultant metric targets provided a pace of progress that enabled achievement of the strategy. Although a top-down approach may seem to be an arguable methodology, the results were never in doubt. The improvement task was woven into tightly linked planning processes covering strategy adjustments, business cycle targets, and organizational development that consistently stretched the enterprise. Significant collective improvements were consistently achieved, in large part because the planning and execution efforts were so tightly integrated and the improvement targets were grounded by their baselines. The involvement and discipline built into the execution and follow-up became the differentiator. The key lesson here is, "If you can visualize a path, you can follow it."

1.3 Communicate the Vision – Inform and Sell to Build Excitement, Engage to Generate Buy-in.

Once it's developed, the final step is to communicate the vision. Although the primary motive is to inform and engage the rest of the organization, it's held until the last step in the process. Pushing an implementation plan forward without preparing the team for the news can have divisive results that might impede buy-in and delay progress.

Inform. Providing the appropriate information is the first step in communicating successfully. Bring the team up to speed on the facts supporting the vision, what's required, and why it's necessary. Most people will buy into a better future, so the vision should target something better. 'Rooting your vision to reality' uses the lens of attainability to build credibility. A vision that entails stretch objectives won't automatically make sense to everyone, but if a competitor has already blazed a trail to similar achievements, the organization's need to meet or improve upon the competitive achievements helps grounds it – regardless of how far it stretches them from their current position. A memorable example of this was the message delivered by President Kennedy when he communicated the challenge of the space race during the 1960s. The Soviets had already cracked the atmosphere, establishing a sort of technical precedent; and JFK's timeline was long enough (almost a decade) to minimize the "impossibility factor," making the goal of placing a man on the moon seem plausible. The key reason for doing the groundwork and for taking care in structuring and staging the communication is to accomplish buy-in. The less plausible the objective, the stronger the roots will have to be.

Relate. The next step in communicating the vision is to sell it and build excitement, using two levels of communication to achieve favorable impact. The first and most important level explains how it will positively impact individual concerns: employees need to see what's in it for them. The message needs to be inclusive enough to apply to everyone, yet diverse enough to address each functional group in the organization. The second level covers the favorable benefits to the business itself. It's wise to translate the implementation plan into a favorable benefit for each specific area of the business, and then package them to show the benefits to the overall business. This helps to cover both the "what's in it for me" as well as "what's in it for all of us" components.

Involve. The final step for engaging people in the vision is to describe how they will be affected. Helping them understand how their input will shape both the results and outcome, as well as where the vision might be adjusted to accommodate their inclusion, aids in the accumulation of credibility. Make these descriptions general enough to touch everyone by first, stopping them at the functional objective level, and separating them from the more specific group and individual objectives. This degree of "trickle down" is adequate for this stage of the process.

Focus. It takes all three elements of communication to get everyone focused on the vision. As the communication process is initiated, there are certain pitfalls to avoid. First, never assume that just because a message has been delivered, it has been comprehended. To achieve understanding, messages must be simple, direct, genuine, and confined to essential elements. Doing so reduces tension and relieves anxiety. Second, the communications must be timely. Get word out well in advance of initial start-ups so that no one is caught off-guard or made to feel out of the loop. Staying ahead of the rumor mill will enhance the credibility of the communication. The messages must state the game plan directly, delineate the timetable, and identify the benefits to everyone involved (management, employees, customers, stockholders). Building organizational commitment to the strategy and its implementation are key elements of success; enablers for the leadership team to walk the talk and keep the path to success less steep, making it easier for associates to follow.

Continuity. Implementing the new vision requires adjusting the list of tactical programs that are currently consuming resources. As the transitional strategies are identified, all existing tactical activities that can support the new vision should be retained, while new tactical requirements should be identified, initiated, and monitored for their contribution to success. The sustained activities help keep the amount of change more manageable, easing adaptability and strengthening morale. All tactical programs that no longer apply to the strategy must be deferred or discontinued in order to better focus the team on the new direction, a difficult but essential step to free up resources. Having integrated both sets of actions, timing can be applied to the tactical objective set.

In the 1400's Niccolo Machiavelli wrote, "It ought to be remembered that there is nothing more difficult to take in hand, more perilous to conduct, or more uncertain in its success, than to take the lead in the introduction of a new order of things." He was right, and although Machiavelli's mandates, as recorded in *The Prince*, are often blunt, harsh, and uncompromising, his words quite often have application to 21st century business matters, for he adds, "The innovator has for enemies all those who have done well under the old conditions, and lukewarm defenders in those who will do well under the new."

The modern day interpretation of Machiavelli's words intimates that in order to reduce resistance to change, it is necessary to win over as many colleagues and associates as possible to the potential benefits the changes will provide. Otherwise, folks will maintain comfort with what they already know, the normal routine; or simply put: same "stuff" different day! Clearly stating the expected results, outlining the methods to be used, and ensuring their involvement in the change process will go a long way toward reducing the fear of the unknown, allowing support to build and generating success. These are the steps that sharpen your hallucination into a clear vision.

In the coming chapters, we'll explore ways to put the new strategy to work so as to generate results quickly and maximize efforts. In the meantime, it's important to remember that adjustments to your leadership approach are required in order to facilitate the development, deployment, and execution of a winning strategy for the business. It is ultimately the executive leader's "Response-Ability" to show the team the way.

Chapter 2:
Leadership Response-Ability
Setting the tone for a more Responsive Organization

1) Align for Response
2) Securing a "Confession of Reality"
3) Motivating the Organization
4) Work Enterprise-Wide
5) Design Your Processes
6) Creating Organizational Focus

The entire organization had stepped up to the challenge of containing our critical business issues, and we were starting to see some results. After only three months our largest customers were taking notice of our improvements in delivery and response, and several of them had started to increase their orders, though still a bit wary about the sustainability of our renewed performance. Indeed, many of our internal corporate problems had been contained or solved, but a handful of noisome issues still remained that would require more extensive work.

We had earned enough breathing room on the day-to-day issues to allow us to elevate our line of sight toward finding more strategic solutions to some of our greater challenges. I invited my senior staff and their direct reports to an offsite strategy review. In preparation, they were asked to structure all of the available market intelligence, product and service development priorities, and operating performance information into a review format for the team.

"First, I'd like to make everyone aware that Valhalla Corporation has

removed us from probationary status and rewarded us with two new orders."
The room filled with applause.

"Although I know you all realize how beneficial that is, so far we've accomplished it by lifting ourselves up by our own bootstraps. Our results have improved more by muscle than through finesse. That isn't a recipe for sustainability, so we need to develop longer term solutions. As we discuss strategy for the first time together today, I want you to think about our options for turning the business from survival to growth. For now, think about issues and processes rather than actions. Believe me, actions will come later!"

Each of the functional leaders and their direct reports were asked to present their business information to the greater team. Once all of the presentations were completed, we broke into four smaller groups, each led by a staff member along with an independent facilitator. Their assignment was to develop a SWOT analysis (Strengths / Weaknesses / Opportunities / Threats). When the SWOTs were completed the team leaders presented to the group for discussion, as a starting point to develop consensus on priority ranking, as well as to identify potential countermeasures.

The completed strategy review and SWOTs provided us with ground level information (a combination of market supported facts and experiential opinions) to form the new direction for the business. We still needed to blend that list with the key elements of our parent corporations' strategy to build a set of organizational objectives and set the stage for their dissemination into short and long range goals. We adjourned the meeting to give the senior staff time to construct a master list of potential strategic initiatives based upon the SWOTs, a priority assessment, some suggested countermeasures, and the corporate initiatives lists.

For our second session, we expanded the team to include key next-level functional, technical, and customer leaders. Their mission was to translate the strategic initiatives identified by the senior staff into a more detailed set of attainment projects. The process required that nothing be held sacred -- not previous performance paradigms or constraining business processes; all would be subject to change once they made their way to the final set of priorities.

"Good morning!" I was enthusiastic as I addressed the team. "Welcome to the second round of our strategic development process. By the end of today, we'll have put some definition to a new business strategy. It will be comprehensive, grounded, plainly stated, and supported by a complete list of projects and metrics by which to gauge our progress. When I say comprehensive, I mean that our number one initiative will be customer-facing, while our number two initiative will address improvements to our own business culture. Following those in the order of greatest impact will be operational performance, product initiatives, business process, and financial goals. Once we have defined it all, we will use a process called Strategy Deployment to put it into action as soon as we can communicate it to the organization."

"How much of it has been dictated to us?" a senior engineering associate asked.

"Honestly, there are a couple of corporate projects and some ending financial metrics that are provided for us, but the rest is for us to determine. The staff and I took an initial stab at it last month, so your role will be to help us finalize an achievable plan."

Each staff member was asked to unveil his or her list as the "champion" for that particular initiative. As we unveiled the draft set of organizational initiatives to the team, I could almost feel the sense of relief in the newcomers to the group. It seemed to be a clear indication that the staffs' findings were very close to what the consensus believed would be necessary to turn the business around. We agreed publicly that the options being presented should be considered preliminary and that the specific purpose of the meeting was to fine-tune the list. Their participation and contributions were encouraged and required. We stressed that there were to be no boundaries for the solutions at this time, they needed them to dream a bit.

"Now, I'd like to lead a discussion about our customer relationships. Let's review the data we have from every customer touch-point, including direct metrics and tangible feedback. As you do this, please try to put yourselves in our customer's shoes relative to how we perform."

I paused a moment, then added, "I want to talk to you about Response-Ability. Not the one-word kind." A slide with the crossed out word "responsibility" was placed on the screen. "Rather, my version looks like this." A second slide with "Response-Ability" popped up in front of them. "I use this version to convey that it's my duty as a leader of this business to be responsive to customer needs within the same timing that those needs are expected. My

Lowell J. Puls

job is to maintain these relationships so that they are constantly functional and continuously improving. This is *my* Response-Ability."

We embarked on a lengthy discussion of our present customer relationships, with the customer-facing functions helping set the stage by communicating some facts along with direct feedback about our performance. I reminded them of the need for us to maintain a clear "line of sight" on our both the current and future customer needs by building them in as a goal for each organizational initiative. That approach would keep our goals from being overly internalized.

"Now that we have explored the customer perspective from our own view – what we think they think; I'm going to ask you to take a quick lunch break. After lunch we'll have Paul Johnson, CEO of Valhalla Corporation, give us his perspective on our performance." The group quickly dispersed.

When we reconvened, I introduced Mr. Johnson.

"Greetings, everyone, it's an honor to be here" he began. "When I was initially asked to come and speak to you, I was concerned about how negative the message was going to be. To be quite frank, you were doing so poorly that we were planning to replace you as one of our key suppliers. When we had benchmarked you against other market competitors, neither your ability to supply nor the quality of the incoming products measured up our to standard. Because your performance had begun to impact our own business in such a negative way, we felt that you were leaving us no choice but to seek alternative supply. Although I can say that your recent progress is encouraging, you should also be aware that we can't yet trust that you've fully corrected all of the problems, so we'll continue to explore other supply options on a parallel path"

He carefully articulated his view of our performance, the areas where he felt we needed to focus our efforts, and the level of progress needed to maintain our present level of business. I was somewhat comforted to hear that his impressions of what we needed to accomplish were largely in line with those we had identified in our own discussions.

"How much time do we have to fix it?" Dave asked the perfect setup question, probably because they knew each other so well.

"I would say you will need to be a level just below world-class performance within the next year, and recognize that the bar is constantly being raised. I should also tell you now that the penalty for failure will be extreme."

His timeline was much shorter than we had imagined, and I could sense that his more dire prediction of consequence was somewhat unnerving to many of those in the room. He completed his remarks and followed with some casual Q & A before departing. The discussion had helped greatly in setting the tone for the balance of the day's efforts.

With a more urgent perspective of our key customer's point of view, the team then divided into cross-functional working groups with the assignment to draft the game plan for implementing the initiatives, thus turning the vision into reality. The deliverables for this exercise included a description of the objective, a clear definition of the metrics of success, the identification of an initiative champion and functional ownership, and, finally, a list of potential projects. Once they completed this step on one initiative, the teams rotated topics to pick up where a prior team had left off – building upon what that earlier team had accomplished. They were encouraged to work cooperatively with the teams ahead of them to resolve or elevate differences in opinion on any of the proposed solutions. I continued the process sequence until each team had been given a chance to review and provide input regarding each goal.

By this point, the benefits of the session were beginning to become more obvious to the team. Collaboration was already improving, and it was providing solutions that had a more cross-functional flair. The objectives and projects they had identified and prioritized were capable of delivering rapid improvement. Also, a broadened sense of commitment to deliver the improved results was developing. Much of the confidence in attainability that was building emanated from matching the objectives and metric targets to the market needs. Each participant was starting to envision a better result.

"I truly appreciate all of your engagement and participation in this process. The next step will be to communicate a top-level version of the strategy that is understandable to the entire organization. Then we'll solicit their support in finalizing the project team assignments before kicking off the work." After a few questions, we closed the session.

Since each individual in the room was responsible for managing a segment of the organization that collectively added up to its entirety, I asked them to consider how they would allocate resources to the programs as well as solve the need for project leadership, transition for sustained projects, and an approach to eliminating noncritical programs. These would all have to be completed in our next session before communicating the strategy.

* * *

Though the term 'Leadership Response-Ability' was introduced in the opening chapter, the concept deserves due explanation . . . at least a chapter's worth! I wrote an article in 2010 defining it as providing leadership that's responsive to the needs of an organization, with timing and an intensity that satisfies those needs. Every market has competitive forces that define the response requirements in which its participants must provide products or services to its customer base. A providers "response model" must be designed to meet the markets requirements in a time frame that makes him competitive in that market. Stated more plainly, if you sell products from a shelf, then you must be able to replace the product as quickly as one sells. If you manufacture products to order and advertise an eight-week order fulfillment time, then your materials and manufacturing system design must enable an eight week response even when demand changes. Because this timing is market driven, it necessarily applies to every peripheral support activity that affects fulfillment: decision timing, organizational objectives, and metrics management. Fulfilling your "Leadership Response-Ability" requires you to incorporate this timing into your management review activities in order to maintain their cadence with the needs of the market and the business, developing a culture of responsiveness.

Align for Response

But how do you establish this type of responsiveness in an organization that doesn't possess it? First, you have to determine the timing requirements present both in the market and in the business. They may be different depending upon the complexity of the business itself (multiple markets, etc.). The starting point is to define *takt*, a German word for rhythm that in *Lean Enterprise* refers to the "rhythm of the market." Although application of *takt* has usually been reserved for the manufacture of end item goods or services, it can also be used to define the cadence that applies to business processes and that will enable you to utilize the rhythm of your own market to drive all of your business activities, better linking them to results. Aligning everything you do around the cadence of a businesses end deliverable is just a smart thing to do.

Next you should build that timing (or segments of that timing) into your business processes. Your review cycles play an integral role in this, because reviewing a metric too infrequently for you to take corrective action will misalign the resulting actions from the business and market results, causing performance erosion. A great place to start is with your program and metrics reviews. These typically have some discipline already associated with them and it's easier to tie their timing to expectations of behavior.

Finally, you'll need to introduce sustaining methodologies to help you stick to the disciplines, refine them and constantly confirm their alignment with the customer base. We'll take a more in-depth at these in chapters 6 and 10, but for now, you'll want to perform a deployment check in order to maximize the involvement of your organization. Last, add a periodic improvement review to turn process performance information into refining actions.

To be truly effective in driving the implementation, be sure to utilize your leadership teams understanding of the organizations culture. You can speed deployment by personalizing the pace and content of your communications, increasing engagement and ultimately, your success

Securing a "Confession of Reality"

Engage your leadership into understanding the organization, and then translate it into targeted action.

As a leader, it's always helpful to remember that patience and impatience both are essential components of your management style. In saying this, I'll stress that patience is the foremost characteristic that will encourage people to follow you, but the judicious use of impatience is a fundamental leveraging tool for the call to action. While both are essential, my belief is that striking a balance slightly in favor of patience will allow people to get more comfortable with your management style and as a result commit to your change initiative.

Steering the leadership team toward a confession of reality takes more patience. When you're trying to get team leaders to admit to the current state of the business, regardless of what it is or more importantly, who is

23

responsible, using impatience will only scare them into a false reading. Self-admission is powerful tool for generating buy-in both to the problem and the solutions, so the creation of a mutual reality (the businesses and their own) is an important enabler for people to understand the need for change.

You should always try and ascertain an organization's culture from two perspectives – one by observing its embedded behaviors and the other by assessing the internal and external influences that tend to move it. Let's look at embedded behaviors first.

Identify Embedded Behaviors

The first behavioral trait I'll want to assess is organizational focus, which is simply about what they're actually working on at the time your change initiative is beginning. It's best to just walk around and ask people what they spend most of their time doing. Don't look for a detailed explanation as much as simply a general idea of the type of activities they're engaged in. A complex answer is much better than an "I don't know," but either extreme is actually bad. Lack of focus will strand work in-process and short of completion, while intense mono-focus can often leave necessary activities unattended. I once had responsibility for a facility whose culture was described by an outside observer as "ants at a picnic!" He was telling me that any time an issue popped up, everyone ran to it. Unfortunately, that behavior occurred without real priority, causing numerous important projects to be interrupted or dropped, often preventing them from taking problems to closure.

The second critical behavioral attribute is responsiveness. Culturally, this can range from either an organization that puts its customers first (very good), to one which looks at its customer expectations with a sort of disdain, making excuses as to why personal issues or business internal matters are more important (very bad) than the satisfaction of market needs that they consider inherently wrong. A responsive culture intrinsically understands that whereas its customers aren't necessarily always right, they are always their customers and it is they who keep cash flowing in. As such, they should always take priority over internal issues. I can't tell you how many times I've entered an organization where the internal attitude was that the customers never know what they want, and because of that, "their emergency is not ours," so we'll put them in line and get to them when we can! Unfortunately, being responsive to any customer needs in the timeframe they require is your opportunity, and building a capability that enables you to do it cost effectively can be an integral strategy to win over marginal markets. Some of the most successful businesses have redefined their market through responsiveness! Customers settle matters with their feet—if dissatisfied, they walk out the door and never return!

Another vitally important cultural behavior is adaptability, and it too can be a mixed blessing! Excessive adaptability might create an organization that eschews standardization and makes control difficult to achieve. While all organizations require a certain number of standards, set procedures, and well controlled practices to ensure consistent delivery of products and services, they cannot be so constraining as to rigidify response. An inflexible attitude that limits your ability to accommodate occasional out-of-the-ordinary customer requests will label you as a company that's too rigid to be counted upon in emergencies or too staid to co-develop new product new offerings with. The proper balance point is to have a stabilized structure and delivery processes that are reliable but that also have exception management contingencies for suggestions, alteration requests, and product improvements! Without process stability, there can be no consistency of performance, and without sufficient adaptability there can be no acquisition of new markets or advancement in areas such as research and development, key requisites for growth.

Communications is the final embedded cultural behavior that should be understood up front. Organizations that don't communicate well tend to be functionally isolated, causing a failure to work effectively between departments, ultimately forming "silos." The visible existence of any type of timely, disciplined communications moving between levels and across functions, either through periodic reviews and/or general information sessions, is a great indicator of a company's commitment to communications.

The chart below demonstrates how these four behaviors inter-relate to define a culture, affect customer relationships, and potentially impair the ability to drive change. It further suggests that a balanced culture will most readily accept and adapt to a change initiative.

Benchmarking Culture	Organizational Result				
Customer Satisfaction Level	Low	Satisfied	Highly Satisfied	Satisfied	Low
Cultural Trait					
Organizational Focus	Focus Deficit - many activities without priorities, no discipline to finish in process activities.	Too many priorities, some focus but prioritization is lacking	Focus is organized along key initiative execution, priorities are clear, achievement level is high	Minimal activities or priorities, very focused	Intense mono-focus - very few initiatives in process.
Responsiveness	Ultra High Sense of Urgency Frenetic	High urgency, over reactive	Highly responsive	Responsive when pressed	No urgency - Organization is generally non-responsive
Adaptability	Highly adaptive - nothing is standard, customization is not controlled	Adaptive - few standards, excessive customization	Timely adaptation to market changes - Standards are leveraged and customization is controlled	Limited adaptability, very standardized, customizes when forced	Not adaptive - Everything is standard - nothing is customized
Communications	Excessive communications - Can't keep up or meaningless	Communication level is high, but value is low	Timely-clear communication of org. goals and progress	Some structured communications, value is low	Minimal to no communications - doesn't generate interest (apathy)
Rate of Change	Resistant	Constant	Controlled	Inconsistent	Resistant

Understand your "Spheres of Influence"

The origins of some cultural behaviors can be difficult to discern, but influences that can originate both externally and internally to an organization can play a huge part in defining a culture. Once the embedded cultural behaviors are identified, greater understanding into what drives them can be gained by looking at how influences align themselves to exert force. Three primary factors generate cultural influence, and I've arranged them into spheres, with variable overlaps to depict a specific emphasis. The proper arrangement of these spheres should mimic the way those influences actually affect behaviors in the organization, similar to the way a bubble chart works. Once understood, they can be useful in defining your plans to affect cultural change.

Stakeholders is a word companies use to describe all of the people inside or outside the business who have a vested interest in its performance. To maximize the effectiveness of this definition, let's broaden it to include anyone on the inside or the outside of a business who can materially affect its "Response-Ability" to adapt and perform in its markets. Stakeholders on the outside of the organization can exert influence over the behaviors of those within it in ways that will impact decisions and even performance. Typically, it's through people who have business contact with those on the outside: the CEO, senior executives, a buyer, perhaps even the executive secretary. Let's

KISS: Keep It Simple and Sustainable

try to understand how stakeholder influences can align around functional responsibilities and affect organizational thinking.

To understand the dynamic a bit better, consider spheres of influence to represent factors that affect the ability of each organizational segment to make decisions and/or influence the activities of others, both inside and outside the organization. It's an important concept when it comes to building your stakeholder list. The three spheres include; the social-organizational sphere; the customer sphere; and the regulatory sphere.

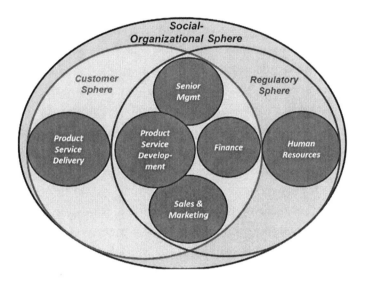

Regulatory Sphere - In the day-to-day operation of many businesses, regulatory influences can often trump the others for their ability to impact decision making and process structure. Much of this is due to the weight of legal accountability that these requirements can impose, as well as the specific mandates that might be scripted within day-to-day operations. Both the number, and the importance, of regulatory spheres will vary by industry.

Customer Spheres – These influences are the second strongest because of the power exerted by the needs of your customers, and the customer-centric behaviors present in many types of organizations. Even so, it isn't uncommon for them to alternate in strength with Social Spheres based on a specific business type or even at different times of the year..

Social Sphere – The influences that form the social group originate internally in the form of company specific (cultural) behaviors, and also externally from the local community. These are the least dominant group by virtue of their relative strength, yet overall are more consistent in their behavior and as a result self-sustaining.

Consider the impact spheres of influence can have on individual business functions when looking at how they are managed on a daily basis. An inordinate strength in one functional sphere (such as regulatory oversight in accounting) can often create an organizational weakness in the other functions if the regulatory responsibilities transcend the true requirements. Excessive approval and audit steps used to assert control can constrict process flow and can dull response for delivery and services. Any function can suffer from an imbalance in influence – human resources, accounting and sales all have regulatory requirements to fulfill; human resources and operations are heavily impacted by the local-social environment; and operations and sales will be influenced by customer requirements and internal stakeholders. Maintaining mutually held, cross-functional objectives with disciplined timing in execution helps to balance task loads and deliverables, and the strategic use of timing can be effective in neutralizing the negative effects of any influence.

Your primary leadership Response-Ability, is to assess and monitor the individual business functions and to help them to maintain a balance between the organization's needs and the individual influences that drive them, thus preserving the best interests of the business from all aspects.

Motivating Response-Ably

Experience suggests there are numerous ways to motivate employees to do quality work within the time available. Among the most common, the metaphorical "burning platform" can be used to ignite engagement through a fear of disaster. A "winning team" approach, similar to cultivating a positive attitude on a sports team, will draw engagement by virtue of its positives. A last example might be referred to as a "benevolence" approach, describing a change initiative that enlists engagement because it's for the good of all stakeholders.

Which one is best? The answer really depends on a lot of things: the current performance of the business, the trajectory of its markets, the intensity of the competitive picture, and the stability within the organization.

A burning platform is primarily effective in times of dire need: an unprofitable business, a declining customer base driven by poor performance, major/sudden changes in the market or the economy. The approach comes with both benefits and limitations. On the benefit side, the immediacy of a burning platform can engender camaraderie of purpose within the organization. The imparted sense of urgency will rise to equal the intensity of the "fire," and can rapidly galvanize the team. The limitations of the approach greater; it must be a real condition, well-communicated, commonly understood by the organization, and credibly supported with actions and data. False alarms will undermine leadership credibility and ruin the effort. Further, any initiatives that arise from it must be focused on a specific set of results and when those results are met, the fire is extinguished. Finally, the time frame for the initiative must be short, for just as a fire burning constantly alters in appearance and unfed will die, so it goes that extreme urgency is not perpetually sustainable.

Motivating a team under the guise of a winners theme can generate sufficient momentum to elevate the business to the next level, but not enough to rescue it from a collapse. Conditions that are viewed as generally positive might favor a winning change initiative, while a severely under-performing business could be viewed as a stretch, breeding doubt and disbelief in both the strategy as well as the integrity of the leadership team.

The use of benevolence as a motivator bears merit, but lacks excitement. It draws its strength from the "what's in it for me" mindset, because of the mutual security that great improvement can generate. It can sometimes be perceived as a diversion – absolving the organization of the responsibility for the need to change under the guise of "someone else needs it." Improperly managed, benevolence commands less urgency, with a slower rate of commitment, and more relaxed buy-in than the others.

While a leader's individual style can play a strong factor in determining which motivational approach to use, the reality is that equal weight must be given to the situation, including both the business condition and the organization's culture. More usefully, the motivations for all associates should be supported by the nature of the drivers for the initiative: failure = crisis management (burning platform), underachievement = turnaround management (winning team), modest success = transition to business excellence (mutual benefit), and more consistent success = market leadership / top-tier performance through business excellence (high-level mutual benefits). If the motivational techniques chosen match the business condition and its needs, associate buy-in becomes a natural function of alignment with the confession of reality.

Work Enterprise-Wide

Working enterprise-wide involves demanding that each aspect of the business attain balanced levels of high performance. Organizations often develop specific excellence in narrow aspects of their business and exploit them to an acceptable level of overall success, but then fail to strengthen the balance of the organization to match the proficiency. Exploiting dominant organizational capabilities can result in a skewed business model that rises quickly on those strengths before falling under the weight of other under-performing elements. Challenging markets, rapidly changing conditions, or a focus on a sole strength may create unhealthy dependencies, inhibit growth, and lead to consistent under-achievement because of an inability to adapt. For example, a company that grows seismically by catering to one precious customer, but does so by compromising its performance with other customers or by de-stabilizing other important company functions such as product design, might experience a catastrophe with any change in that customer relationship.

Start by identifying the inter-relationships between each of the business functions and connect their responsibilities through project assignments and performance metrics. In doing so, you begin to create a bond among them, generating synergies that will improve collaboration and performance. Next, engage everyone on the teams by assigning them roles that support the objectives and also establish broad responsibility for the attainment of key metrics. Failing to fully close the loop with engagement could allow the individual processes to malfunction, eroding performance.

Consider a sales team that is calling all of the shots relative to product offerings and the variety of feature offerings. Suppose further that the organization has fallen into the trap of offering nearly everything conceivable as options on the product, and in the process has impaired its ability to deliver products to customers on time. Because every order had a customized appearance even when assembled from a group of standard offerings, the organization no longer regarded any of its products as standards, and it forced each of them through a customizing step that added weeks to the

KISS: Keep It Simple and Sustainable

completion time, but without adding any real value. In the actual example, the process impeded their delivery response time by as much as 50% and was costing them business.

In an effort to understand the above problem we performed a classical analysis of the product options, their individual demand, and the revenue generation of each. We then teamed the accounting, manufacturing, and engineering functions with the sales force, and tasked them to cooperatively streamline offerings without losing business. The result was an impressive new line of standard product offerings with a focused set of custom options. In some cases they were able to standardize certain options at no cost, gaining profitability from reduced internal processing, while in others we leveraged additional margin from specialized offerings. The solution improved the fulfillment capability of the business while satisfying most critical customer requirements, and the improvement in market responsiveness actually gained us sales in the very first year rather than losing market share due to any reduction in offerings.

Process Design

A company's processes are a critical starting point to facilitate an enterprise-wide approach. When reviewing the designs for essential business processes, it's important to ensure that their inputs and the output deliverables flow through every appropriate function with the correct weighting and timing. Linking your processing time to the market cadence ensures system response, while the multi-functional deployment helps to avoid an overabundance of control or authority by any single function. Carefully structured metrics and timing create flow and support decision-making that fulfills the organizations need to respond.

Let's look at an example of delivery to the customer, a key metric in nearly all businesses. If the delivery metric encompasses being on time to the customers' request, and the business sells a custom engineered product, then achievement of an on-time metric must result from managing a sequence that moves through writing the order, purchasing supply items, following through with the engineering of custom features, and producing the product or service, all within the prescribed market expectations. Too often, a failure to assign time limits causes the fulfillment process to break down and the customer expectation to be missed.

31

This timed connectivity works just like any kind of sequenced activity. On a football field, the slower pulling guard is given a shorter distance to run to his blocking assignment or the faster running back won't have any protection. In a manufacturing plant, each step in the assembly sequence must be balanced to have nearly identical duration. If they don't, the product doesn't flow smoothly, the effort is less productive, and costs are elevated. And, in an office process such as payables, all of the steps in the payment process have to be structured so that the payee is remunerated on time, every time. Referring back to our original example, if the response cycle time for engineering consumes too much of the systems order fulfillment time, the process steps must all be redesigned to be completed within the appropriate *Takt time* (market demand time).

Maintaining enterprise-wide deployment ensures that not only will each department measure their processes using compatible metrics, but also that the processing steps and timing will be linked and designed to provide a continuous flow of activity. Extending the story further, if the processing through engineering is four weeks, procurement another four weeks, and manufacturing takes 18 weeks – accumulating to a total completion time of 26 weeks, in a market that will only tolerate 16 weeks, there is a clear disconnect to resolve. By starting at the customer and translating the market requirements back through the company's processes, they can be reengineered to that baseline, improving your ability to consistently hit the customer or market response requirements.

The last component of working enterprise-wide is organizational focus, and it involves aligning the entire organization's priorities around achievement of the key business objectives.

Creating organizational focus is one of the most difficult aspects of transformational change because of the constant need to manage your engagement in initiatives across all functional lines. It's within the individual functional silos where activities and priorities become de-coupled from the primary goals, diminishing your ability to achieve key initiatives. These activities have to be carefully managed to support functional excellence while at the same time maintaining cross-functional involvement on the key initiatives. Keeping the key initiatives, the higher level metrics and the related local timing requirements in front of people will facilitate consistent prioritization of functional objectives.

Prioritization can be further clarified by triangulating the initiatives among the customer's critical needs, those of the business, and the market requirements. These three points will ground them to reality, while addressing the real challenges in the business. Further, they lend credibility to the vision, which in turn helps people to commit themselves to a more clearly identified change initiative. The factual basis for these priorities enables alignment of the implementation actions into a focused plan that achieves the desired result. Any activities tied to non-critical initiatives should then be placed on hold or ceased completely, to avoid wasted resources, and focusing the effort.

Having narrowed the initiatives list to a manageable number, the deployment plan can be primed by a group evaluation aimed at gaining a sense for its impact to the overall culture, as well as to quantify the individual strengths, weaknesses, technical and interpersonal skills resident in the team. The goal is to unearth two types of associates: 1) those who might inhibit implementation due to lack of either skills or willingness, and 2) those who will facilitate it. The evaluation will identify opportunities for redeployment in order to balance team memberships with appropriate skills, and where to position the strongest leaders in order to leverage your success on the key initiatives.

Strengthening and aligning the organizations cross-functional relationships also helps accelerate decision making. With the vision established, the initiatives to achieve it outlined, and the cross functional assignments defined; the task assignments need to be distributed as evenly as possible throughout the organization. The old adage of "work gravitating to those who get the job done" should be avoided through active management of the deployment. Individual associate evaluations will assist you in balancing the player strength on each team. An associate who might inhibit progress can be neutralized by positioning a stronger person who is capable of managing for success. Finally, it becomes essential to ensure that the vision level metrics are broken into more granular local measures (team or functional level) and aligned to specific projects. Properly identified, these should feed the vision level metrics.

Focus becomes even more sustainable when the plan accounts for the long term involvement required by all associates. Maintaining an enterprise-wide approach supports a more balanced set of business objectives that enables steady improvement. It's this balance that prevents exploitation of the relationships and inter-dependencies inherent to each business function. In essence, the overall group focus has migrated all the way down to the specific associates through assigned actions. We touched on this in chapter one in the sequenced need for understanding the organization, translating the vision into actions, and deploying those actions across the organization, in the process, balancing the level of change across all segments of the team.

Response-Able Planning and Execution Are the Keys to Success.

Business leaders must embody their organization's values, ethics, and mission. As we learned in this chapter, leaders must also practice their responsibilities with the same disciplined sense of timing as is required by its own competitive market. Reinforced by the leadership, this timing can have a transformational influence on organizational performance.

Machiavelli also said: "Make no small plans, for they have not the power to stir men's blood." I've seen that statement hold true time and again. Small plans breed small results, and allow for contentment in underachievement. Since I'm a firm believer that more than 95% of your employees *want* to do a great job, you can embrace your leadership role by positioning them to be fully capable of responding to the opportunities and challenges that lie ahead. How effective you are at sustaining their desire will define your success. Making that vision "visual" to all of the stakeholders who depend on your business, is *your* leadership response-ability.

Chapter 3:
Keeping things RADically Simple
Using the Success Equation; Results = Approach + Deployment

1) "Root" your vision with targeted Results
2) Plan your "Approach" carefully
3) Deploy to engage and inform
4) Use Results to "course correct" your way to continuous improvement

"Now, let's take a moment to talk about the Malcolm Baldrige National Quality Award." The stunned silence in the room bordered on disbelief. "Not because we want to apply for it, or for that matter even win it, but because the application process for that award asks us to look at what we do in the course of running our business in a way that has ultimate simplicity. It's been useful for keeping me focused on success in any initiative. The Baldrige assessment process divides a business into seven areas, then looks at the processes used in each of those areas to ensure that your *approach* to them considers and fulfills the needs of all of the business's stakeholders. Next, it asks us to ensure that your process *deployment* broadly engages all of the potential stakeholders to ensure their needs are fulfilled. Finally, we are asked to define the expected *results* in a way that encompasses all of the needs of the business while engaging its stakeholders.

"We'll start with results, because my experience suggests that the actual set of target results is often predetermined by market benchmarks, corporate edicts, demands from customers, or competitive pressure. Because of this, it makes sense to consider our desired results first, and then determine the approach and deployment required to achieve them. Stated in the modified order, we begin with a grounded, targeted objective (result) that satisfies needs of all potential stakeholders. We will then need to determine how to approach that objective to make it more achievable before deciding who to involve (deployment) to ensure our success. I call this by a formula; R=A+D. Knowing this, I'll now ask everyone to identify the sources for our targeted results. Let's start with leadership. What results should we expect from our leadership processes?"

"What leadership processes do we have?" asked an accounting manager.

"I would think business planning could be called a process," Mary suggested.

"What about strategy? Isn't that a leadership process?" Geoff asked.

"Actually, strategy is a category all its own," I answered.

Mary asked, "Well, what about budgeting and staffing?"

"Certainly both of those are the responsibility of leadership at some point in time, yes," I confirmed, "But let's take another angle at this. Can anyone other than Mary give me an example of a human resources process?"

"Performance appraisals!" blurted one of the engineering leaders.

"Do we do any employee development – if so, that could be one?" the accounting manager injected.

"Thanks a lot!" Mary smiled as she said it. "At least I now have someone to target for this year's appraisal training and development!"

"That will get you a *targeted result*," joked a member of the operations team.

I chuckled, but said, "Okay, now we're getting some traction. Someone please take a shot at the list of stakeholders for an employee development process."

"Wouldn't it be the employees and their supervisors?" Dick offered.

"Are there any others?" I prodded.

"How about the company's customers and owners?" Mary offered.

"How so?"

"Well, say the performance appraisal process delivers improved performance, don't both the customer and the company owners win?"

"Good point. We probably should consider the ways they might benefit from the process. I'm not sure about including them in the deployment group though."

36

KISS: Keep It Simple and Sustainable

I turned on a projector to display the Baldrige list of seven categories on the screen. In order, they included leadership, strategy, customer focus, information management, human resources, process management, and results.

"Using this list as a basis, I'd like all of you to break into groups and quickly arrange our existing processes under each. As you go, try to identify their key stakeholders. This is a somewhat abbreviated way of doing it, but as we begin to put assignments to our strategic initiatives, the stakeholder group must be well represented in the deployment step. That's why I'm asking you to do it this way." They broke into teams and went to work.

The teams dug in and worked hard to select their project lists, assign objectives, and develop a deployment plan that would engage the entire organization. With my continued reminders to keep the plan simple, rooted, and deployable, they did a very good job pulling together a first draft.

"Now that you've identified and categorized your business initiatives, I'd like you to return to square one and recheck the entire product using R=A+D," I challenged.

Judging by the silence hanging in the room, they certainly hadn't seen that one coming. After a few moments an engineering leader asked, "Uh... how, exactly, are we supposed to do that?"

I explained, "Before we proceed on building our strategy deployment matrix, you need to take a couple of steps back and confirm you have a clear understanding that each initiative "directionally" makes sense for the business by the way it supports the results targets. It's just a quick back-check and isn't intended to second guess what we've already done. We'll then assemble them into a draft approach to achieve the results based on what we know now. Finally we can build the deployment team based on the areas affected." I smiled and added, "I know it sounds complicated, but for now we'll keep it at the simplest level possible and avoid excessive detail."

A few people typed away on tablet computers.

I added, "Try to maintain the point of view of those you will be explaining it to when it comes time to deploy it to the individual teams. If we can't explain them with both clarity and an appearance of attainability, they'll need to be reworded."

"But some of these goals look impossible." My engineering team nodded agreement with this comment.

37

Lowell J. Puls

"Results targets can be difficult. If they don't appear achievable, people won't even engage. If they look too easy, then our current behaviors won't alter enough to cause cultural change. We have to give them credible attainment plans, and keep them rooted to the facts in the vision we developed earlier, in order to keep it a reality instead of a hallucination. That's what this entire process will do for us so long as we do the work correctly."

The metrics review only took about an hour because they felt the criteria that had influenced their choice of targeted results were difficult to argue with. Certainly some of the goals appeared to stretch beyond what this team felt it was capable of. Nevertheless, I encouraged them to be left as-is for the time being to see if they could build approach plans that would make them appear more attainable, grounding them closer to reality. Once again, they split into teams and worked through the same methods of refining the approach they would use. Next, they revisited the project list for its ability both to achieve the results and satisfy all of the stakeholder needs.

Finally, we tackled deployment in a similar fashion, confirming who needed to be included and defining the level of involvement, from leadership roles to hands-in activities. We re-examined our approach options to better focus on maintaining broad involvement through the rollout to build engagement and buy-in. Each group was asked to establish follow-up procedures and timing that would utilize as many existing processes as possible, and to define the points at which we would celebrate success.

"As you put on the final touches, I'll want you to keep thinking about how to translate these plans into the words and thoughts of the people you will be presenting them to. This step is critical for you to insure group buy-in. Your explanations should clearly tie a line to how we have rooted stretch objectives to the needs of the business and its associates. It's the only way to build credibility within the entire organization."

Give your vision "roots" with targeted results

As a leader, it can be difficult to comprehend all of the many complexities involved with launching a change initiative when you're under pressure for specific results. This pressure can come as an urgent mandate issued internally, or from outside forces; investors placing profit pressure for example. The common response is to jump ahead and formulate objectives based upon the need without considering how to implement them. When I said earlier that the difference between vision and hallucination was a plan, the reference was to the need for a simplified or high level plan. All large-scale change initiatives will expand into greater detail at each phase of implementation planning, but that level of detail is too complex when viewed from top to bottom and it can't be effectively managed in its entirety. Arranging the detail at the correct organizational and functional level deploys it in a way that maintains continuity, while at the same time keeping it at a manageable size to deal with.

Maintaining reporting that is striated and focused also makes it simple enough for everyone to view and act upon. Since ultimately the lower level details develop into the assignments for those responsible for getting the work done, performing status reviews at the functional level plus one-level-up for normal performance will keep them concise. When performance is in doubt, expand your review oversight to two-levels-up to increase your options for corrective action. This approach eliminates the need for people to maintain detailed knowledge of all programs and helps associates zero in on the necessary corrective actions. At the leadership level, you'll maintain a dual view of the objectives, one focused on the performance of the vision level (global) metrics and another at activities taking place at the initiative level. Everything below this is reviewed between each working group and its manager, but only the global metrics are reviewed publicly. Driving out the multiple levels of detail keeps the visions initiatives more understandable for all, guaranteeing their sustainability.

In staying true to the theme of keeping it simple to make it sustainable, the leadership task is to facilitate a basic level of organization-wide understanding for the top-level metrics. Simplifying this information supports the ability for everyone to engage in planning and communication activities that satisfy both functional and personal needs while enabling a broader understanding of how each individual's contributions accumulate into the visions attainment.

Although results may not seem like the natural starting point to frame an initiative, there are many forces that combine to make it appropriate more often than not. The pressure exerted from influences such as competition, customer requirements, regulatory impacts, technological advances, shareholder expectations, financial pressure from lending institutions, and internal employee needs will impose objective targets for you. These targets

Lowell J. Puls

will drive business strategy and initiatives directly, requiring a response that can also serve as your call to action. Beneficially, their external nature will also ensure that your objectives aren't constrained by internalized thinking, allowing them to function as the roots of your vision. In many cases an urgent need for response can singlehandedly "root" a stretch objective, mandating its attainment.

Targeted results also serve to define reasonable expectations for an effort. What kind of return can be expected for the business in the form of process improvements? How can you insure that those results will support the fulfillment of the vision, and how can they be sustained?

This chapter was titled R=A+D to identify the power that external and internal drivers have to predetermine a set of required results and the influence they can exert on how you approach the problem. Without this, your objectives might be defined solely by internal pressures and traditional thinking, ignoring the need to alter performance in the eyes of your customers and stakeholders. Managing a business to internally driven results targets is a common pitfall that can leave substandard performance at the customer unaddressed. A former boss of mine referred to this as "drinking your own bathwater." You have to honestly believe that basing your objective set primarily on the needs of the market, the owners, customers, and employees, is the right thing to do, and that the needs of the business will be satisfied through the correct strategy. Otherwise, if your objective targets become too internalized, your "drink of the day" could be a bathwater cocktail.

Using the appropriate external and internal information to develop target metrics will frame the vision, and your final metrics can be dialed into based on what's needed, wanted, or feasible. This maintains a set of results that 1) satisfies the needs of stakeholders, 2) is meaningful, 3) can be clearly communicated, and 4) is broadly deployable. Over measurement comes at the expense of diluted focus, and insuring that a minimal set of key metrics are clearly defined keeps them more readily attainable. To guarantee a thorough and effective, but simple impact on the organization, focus your metrics on five key areas, plus a sixth factor of time. By priority, the general categories recommended are:

1. **Safety (or) Wellness: Employee-Focused** – While manufacturing companies are often "Safety First," other companies should consider putting employee well-being up front.

KISS: Keep It Simple and Sustainable

2. **Quality: Customer Focused.** External customer "touching" first with a second level that is internal-process focused.
3. **Delivery: Customer Focused.** Delivery of products and services that meet customer needs within market defined timing.
4. **Cost: External Focused.** The total cost of delivered products or services to the customer.
5. **Cost: Internal Focused.** A key metric for cost of operating performance
6. **Time:** A business response metric that is market defined and applies to the entire enterprise.

We'll examine different aspects of this results list later before diving more deeply into how to prioritize and manage them in chapter 8. For now I'll just stress that it's equally critical to insure that each result is measured in a meaningful and accurate way that is representative of how the customer of that metric wants to see it. This helps you to keep all metrics clearly defined, easy to understand, and targeted on achievements that will transform the performance of your business. Consistency and accuracy in measurement are a key element for getting people to trust and gain confidence in the metrics.

Plan your approach carefully

Once your results targets have been agreed upon, it's time to define the initiatives required to achieve them. Although the targets themselves may in part dictate your approach, the way they have been triangulated from past performance, competitor positions, and customer needs will serve to shape the necessary actions. Your choice of attainment tactics must be equal to the significance of the challenge, and sufficient to bridge the gap from your current level of performance to the new target. If you continually confirm the targets attainability, and engage your employees in refining the objectives, the process will build credibility through its support for the varied interests of all stakeholders.

Let's take a look at the example of market share, where setting your objective based on a simplistic desire to "increase market share," might result in a set of actions with right amount of improvement potential. Alternatively, setting an objective based on a market developed target could redefine the

41

Lowell J. Puls

objective to read "regain market leadership by closing the current 20-point share gap between ourselves and competitor A." The latter objective is more clearly defined and should force a more refined game plan for its achievement. This example closely follows an oft-reinforced lesson: a stretched objective will return a better-than-expected result, whereas a conservative objective usually struggles for both definition and results. The use of external benchmarks (the 20 point gap) to "root" the target, before shaping your approach to achieve it, can dramatically increase your odds of success.

Begin filling in the details of your organizations initiatives by asking associates to view the importance of the objectives from the perspective of the remaining stakeholders. You'll need their participation to define the specific projects that will satisfy their needs, uncover any necessary information, establish the lower level metrics, and possibly even identify the need for external resources. The intensity of the approach needs to be satisfied in the deployment stage by assigning appropriate resources to match each initiatives size and priority.

Returning to the example of increasing market share, a target for 20 points of share improvement will likely dictate the initiatives which are capable of achieving the objective. They might include new sales promotions, key product launches, changes in pricing, changes in order fulfillment response time, even monitoring the timing and response to external forces such as regulatory and technology changes. By ranking the list of options available, the methodology can be constructed based on projected impact, deriving a natural priority. Continual cross-checking for project contribution will further ensure that the approach fully supports achievement of the end goals for the business.

Deploy to engage and inform

Recruiting the engagement of your associates requires broad deployment to those who are capable of driving improvements, use and refine processes, and who will mutually benefit from the effort. Having assessed the skills and resources required for each strategic project, a more balanced deployment will come from involving a cross-section of all of the potential stakeholders, and increase your chances of success.

Deployment efforts should have three key objectives; 1) protection of stakeholder interests, 2) contribution to initiative success, and 3) broad involvement of the organization. Engagement helps connect associates directly to the initiatives through their participation in task-related personal goals and in doing so, gives them a stake in the outcome. Participation further helps them to understand how their individual goal performance drives the project and how these contributions accumulate to achievement of the strategy. Hands-on involvement in uncovering the "roots" helps to link any goal "stretch" closer to reality. In many cases, deployment can extend to anyone touched by the initiative, while in others the size of the group can be reduced to only those required for implementation. Involvement might even extend outward to customers and other stakeholders. However it is quantified, engaging everyone who will play a significant role in the implementation or sustainment of the specific goal enhances your odds of success.

A correctly structured deployment plan has two components. First, it requires a finalized project list with all non-priority projects cast aside. Second, both of the afore-mentioned structure and staffing evaluations should be completed so that all limitations in skills and resources become known and can be adjusted for. The deliverables - a clean task list, skills gaps resolved, balanced resources assigned to each project, and strategic placement of developing leadership - are the requisites for good deployment.

Strengthen your deployment effort from the beginning by assigning the team leadership roles to the best and the brightest in the organization, those employees with either natural or experiential leadership skills. Since all team structures should be cross-functional, the functional origin of the potential leaders isn't always critical, unless the project requires specific technical skills or regulatory knowledge. In making your assignments, don't allow technical skills to trump leadership skills, as the latter is far more important. Each functional leader should be assigned a mentor – best accomplished by the senior staff champion for the top-level initiative.

Once these leadership roles are established, each team should be equipped with the proper technical skills, dispersed through the cross functional assignments. Use the staffing evaluation to balance the strength of the players. As an example, it's wise to balance a potentially problematic associate with another that is oppositely strong. Though there are many reasons for retaining a "problem-child" on a given team – usually critical experience or key technical expertise - the presence of too many problem associates might indicate the need for a different kind of cleanup effort.

When planning your deployment, observe some basic rules. Try to avoid excessive multiple team assignments, especially for the leaders. If a leader has been rated as developmental, bolster them with a good mentor, a good development plan, and a stronger team. If there are new employees, make sure to place them in a group where their leader and the associates assigned with them can help them to acclimate and flourish rather than excluding them due to a lack of internal knowledge. Sticking to these rules will absolutely yield both a better change initiative and a better result.

Use Results as a guide to course correct your way to continuous improvement

Setting results targets up front or, establishing the "R," supports development of a good approach and formalized deployment needs that are realistic and aligned to those targets. With both the A and D identified, your results can be monitored and adjustments made to stay on track, or to revise objective targets up or down based on the progress achieved.

In all three -- results, approach, and deployment -- a feedback and improve cycle will add structure to the process and facilitate consistent and predictable improvements. The Results / Approach / Deployment methodology serves as a great sanity check to insure that the methods used in defining the business objectives also support broad involvement in both the implementation and measurement of success. The engagement that builds from executing the process in this fashion will meet the fundamental needs for success in any change initiative.

Summary

So, how do the results, approach, and deployment simplify the process? Because remaining diligent about asking that same series of questions for all actions undertaken, will maintain a clearer focus on both the objectives and the path to success. The simplicity of R=A+D is an enabler for an organization to target, involve, and reward key stakeholders, providing a simpler way to "back-check" and avoid confusion. By using the instructive logic of the Baldrige methodology, to confirm that each objective has passed through the logical sequence of targeting Results, defining the Approach, and using them to build the Deployment plan, you will maintain the clarity of the objectives in the eyes of those tasked with their implementation. Prepare your communication plans carefully (to avoid unwanted surprises and negative perception as the initiative is rolled out) and quickly publicize the achievements to accelerate the plans credibility.

The deployment step in the process also supports a sustainable method of ensuring organizational engagement, breaking the top level objectives down into tasks while maintaining linkage throughout the organization. It creates a direct connection from each employee's individual goals to their daily work, a critical step that's often poorly executed. Be aware of deployment that's too thin or insufficiently cross-functional, for it can break the process down and cause under-achievement. The organization-wide participation must be comprehensive, and the engagement must be continually monitored.

Chapter 4 will look at how to prepare your organization for change, both physically and mentally. Of course, Machiavelli had something to say on this topic as well! "Whosoever desires constant success must change his conduct with the times"

Chapter 4:
Organizing for Change
Align Your Structure with High Performance Staffing

1) Isaac Newton's Laws of - Change?
2) Preparing for Change
3) Building Engagement
4) Measure Impact!

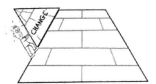

As the scope of our change initiative began to take shape, it became clear to us that the projects would impose resource requirements that the existing organization couldn't provide. In order to resolve the problem, we scheduled a session to assess the readiness of the organization's resources to adopt and implement the strategy. Having taken a two-week break following completion of our strategic initiatives, the team reconvened to an offsite location so that we could better assess the organization and develop a plan to close those resource gaps.

I explained my motives to the group, "So, we now have some definition to our overall strategy and have identified the initiatives and projects that will enable us to accomplish it. The next step is for us to look at our organization and its people from two perspectives. First, we're going to look at the skills present in the organization to see if we have everything we need – we'll call this structure. Second, we'll look at the depth of our human resources to confirm that our capacity is sufficient for us to move forward."

We began with a review of the strategy and the initiatives list. Our purpose was to focus on those initiatives that might demand special capabilities, exposing gaps in our existing skillsets. The staff members had been asked to assemble a skills inventory for their individual functional teams, which we then drafted into an assessment of the entire organization.

After reviewing the initiative requirements, we then covered the skills assessments to focus on the gaps, looking for options to fill them internally with people who either possessed the skills or with ones who had demonstrated developmental potential. In the process we identified the need for an engineering technical specialist on our product development team as well as an internet marketing specialist. The specialized skills inherent to these positions would prevent us from filling them from within and needed to be resolved before progress could be made in a couple of other related areas. We finished this step by prioritizing a plan to close these gaps.

The next task was to estimate whether the organization possessed adequate resource capacity to tackle the complete list of business-critical initiatives and projects. Coming out of the strategy sessions, we had taken the implementation plan down to the project level so that we could make some directional estimates of the resources required. Understanding that precision was nearly impossible we left provisions to abandon certain lower priority projects or to involve outside resources where the payoff merited. The final resource estimates would be confirmed in the project team's kickoff meetings and revisited in the initial group review session.

"Well, the first part was easy. If you need a rocket scientist to build a rocket and you don't have one, it's obvious you'll need to go out and hire one. It's either that or not build the rocket." I resumed the exploratory conversation. "Now let's talk about resource capacity."

"Certainly, we need more resources than we can afford in order to keep the pace we've identified in our plans," Mary observed. "So, we now know we need to put a bell on the cat. The question is, who's going to volunteer to take on that task? How do we plan to acquire the resources we aren't budgeted to pay for?"

I smiled grimly and nodded my understanding. "Belt tightening is never easy. It's going to come down to a matter of priorities. If we agree that a certain resource is vital for the fiscal recovery of our overall operation, we'll have to obtain that resource even if it means having to reduce an effort elsewhere. This is triage, folks. You don't worry about bruises, scrapes, and cuts when there are puncture wounds elsewhere. Make no mistake about it, we are bleeding profusely in certain parts of the body of this organization, so we

KISS: Keep It Simple and Sustainable

must stem the losses in those areas or the whole body will die. First things first." I gave that a moment to sink in and then I said, "Okay, Dick, since cash flow is the life's blood of any business, we'll hear from you first about our company's sales potential."

Dick stayed seated, shuffled some papers, cleared his throat and began. "Well, from the perspective of pure capacity, we desperately need a second sales manager for the West Coast. Our competitors are killing us out there. There's plenty of business to be gained, but we just don't have the personnel out there to capitalize on all the opportunities."

"Do we have good distribution there?" I asked.

"We've got a presence," Dick said flatly, "but it isn't great and that's part of the problem."

"Is the market size adequate?" I was fairly sure I knew the answer, but was steering him a bit. "How does the market size and revenue generation potential compare to some of our higher-performing territories?"

"Let's not talk apples versus oranges," said Dick. "West Coast doesn't compare to the other territories because they buy at an entirely different end of the market. However, we do have a vast range of products we could be selling out there – things those businesses need and have the capital to purchase. The problem is, as I said before, we have no one out there knocking on doors and writing up orders. We've been missing out on this growth potential for years. It's as if we've sent one person out there with a tablespoon and asked him to pull in the waterfall."

"You sound pretty certain of yourself," I said.

"It's all basic research," said Dick. He lifted a page and shoved it across the table at me. "Here's the report I filed six months ago. It verifies the market size, the contracted distribution, and the phenomenal potential for upside growth. I've said all along that an initial investment in the West Coast will pay off handsomely. My pleas have fallen on deaf ears."

I examined the paper. He'd done his homework. The numbers were impressive.

"This is an example of the kind of breakthrough action we need to help us hit our objectives," I said. "Okay, Dick, we're listening to you now. We'll add this proposal to the draft plan. I'll have our folks in Accounting confirm your projections, and then we'll take final action."

I could see by Dick's facial expression that he was equally surprised and pleased. Others around the table also responded pleasantly. It was slowly becoming obvious that the leadership team had some good strategies for turning things around, and it also was clear that these folks were going to appreciate being taken seriously for a change.

The structure and capacity deliverables were necessary for us to assess the team's physical ability to deliver the strategy. Through the course of the discussion, we were able to identify nearly all of the skills gaps, and we gained a rough understanding of the areas where resource adjustments or program timing would need to be reevaluated.

The last step in the process was to discuss areas where our functional emphasis might not allow for appropriate focus or be able to support the momentum requirements of the strategy.

"All right," I said to the team, "let's now see if we can identify any areas that aren't in position to support the speed of project implementation."

"I know that my HR team will have difficulty with all of this," Mary stated. "We're rolling out a new corporate benefits program. The follow-up requirements of the results of our employee survey get more stringent every year, and we've got that new corporate-mandated safety program as well. Diverting key members of my team to support these initiatives cross-functionally is going to be a problem."

"My accounting team is already buried as well!" Jolene chimed in. "There are new financial sub-systems being implemented, and the monthly closing process time is going to be reduced by yet another day. I just don't know how to get it done!"

Dick added, "Our customer service people aren't equipped to deal with the number of customer 'touches' we currently have. If you also want them to chase down forecasting information at the same time they need to follow up on order satisfaction, they're going to be spread too thin. It just won't happen."

I shook my head in disagreement. "You aren't grasping the bigger picture. Keep in mind that the initiatives we've identified will reduce or even eliminate some of the rework-type activities that are currently chewing up your teams' day. The process is going to be streamlined. I'm asking you to 'spend ahead' a bit and commit. The cross-functional team structure will help maintain program momentum when any one group like accounting gets bogged down. If necessary, we'll adjust our expectations according to

KISS: Keep It Simple and Sustainable

the rate of progress and hold weekly barrier meetings to keep obstructions from slowing our progress."

I knew from experience that these team structure and process issues were relatively easy to spot from the inside. Using a nominal amount of process data (including snapshots of the receivables and payables cycles, purchase order approval timing, and other performance information for some of the key processes), we were able to identify some early trouble spots in customer service and accounting. From there we put together a containment plan that could quickly resolve them with resource re-allocation and/ or reporting changes.

The end products of this structure offsite became a shopping list for a specific set of skills and a reorganization plan that would provide an optimized ability to achieve our strategy. Before finalizing it, we held a second offsite session to assess staffing and identify leadership candidates, high-potential associates, and those who presented performance concerns. Again, this followed the structure session by a couple of weeks so the leadership team would have time to both recover as well as prepare. This time we simply conducted a 360^0 roundtable to discuss each functional organization. The senior manager described each individual's performance history, followed by a group discussion.

"Now let's start our performance discussions with HR. Who better to set an example on how to assess and manage performance than the people who help develop our evaluation standards." I almost made Mary wince with that one, but she took the challenge.

"I'll start with Gwen, my HR Labor Manager. From my view, she is bright, articulate, driven, and responsive. The quality of her work is excellent, and she has no issues regarding attendance or personal problems. In other words, I would rate her promotable into my position within a year." Mary paused, indicating it was up to me to finish this report..

"Would anyone like to comment on Gwen?" I asked. There were no takers. "Okay, then, based on my own early interactions with her, I would tend to agree that she is someone with genuine leadership potential. Anything I've asked of her has been completed promptly and correctly."

Geoff had heard enough. "Maybe that's why I can't get anything out of her," he piped up. "Whenever I ask her for something, there is always a reason I can't get it. Plus, she is often rude to my supervisors."

"I would agree with that," Janet added. "It doesn't matter what I ask her for. I just don't get it."

Mary wasn't happy. "I guess it would have been nice to know this earlier."

"Maybe we've just been focused on getting her to change one on one," Dick said apologetically.

"Okay, let's keep things going by doing this. Mary will talk to each of you and give Gwen some feedback based on her performance with her internal customers. I'll let Mary decide if the problems are serious enough to work into her performance evaluation, or if an informal improvement plan will do the trick."

We continued through each staff member in a similar manner, limiting the time we spent so that we could finish in a single day. Once again, the product was an action plan to address anyone needing a development or improvement plan. When we had completed the staffing review, we returned to the structural gap analysis to reassess whether the higher-performing internal candidates could be reassigned to fill the key skills and capacity gaps or whether it would be necessary for us to recruit from outside the company.

In the ensuing weeks we began to implement the organizational adjustments that had been identified. To minimize nervousness, we started with the positive moves, publicizing them broadly to encourage everyone to engage in them. Next, we tackled improvement plans for those associates with developmental or performance concerns, handling them behind the scenes and tasking them individually to improve their level of contribution. As we moved forward, it was clear that our approach, communication, investment in preparation, and broad involvement in developing the initiative plans, metrics, and teams was paying off in greater employee involvement. The organization seemed more at ease with the changes being made, giving us positive inertia.

* * *

While participating in a senior leadership event for a major industrial company, I was invited into a group discussion on culture with the CEO. During the course of that discussion, he made a statement that has never left me. He said, "Culture is more powerful than strategy. When culture goes up against strategy, it wins ten out of ten times." Now I'm sure one version or another of that statement has found its way into countless management books over the years, and I'm equally certain that I'd heard a different version or two of it before, but perhaps because his recital came at a time when I was in the midst of leading my own major change initiative, it resonated tremendously. Why? Because changes in strategy are <u>never</u> independent of a need for some level of cultural change! The executive challenge is to set into motion those changes that improve upon the organizations weaknesses while at the same time preserving its strengths.

Although I would always refer to this process an evolutionary one, it isn't meant to imply that radical change isn't necessary. In truth, it's often unavoidable. Some types of rapidly infused changes – reorganizations, product line or facility rationalizations, and other abrupt moves that might be required for survival, can also be effective at energizing the "inertia wheel" of change. Let's see if Sir Isaac Newton's famous "Laws" can be used to help us to understand the process mechanically.

Newton's Laws

A truly "strategic" initiative will usually require some type of motivational "inertia" to stimulate its culture to engage in the changes. To help explain how this works in application, I'll use all three of Newton's laws of motion, starting with his first, the law of inertia. To implement a strategy in a culture that's essentially "at rest," the first half of his law applies: "An object at rest will remain at rest unless acted on by an unbalanced force." Your change initiative will require a level of effort greater than the size of the mass (the organization and/or its market) to get the "inertia wheel" turning. In this case the "unbalanced force" becomes the application of an energy that is exerted by the strategy and amplified by the speed and breadth of its deployment. Both are required to energize the organization into motion.

Now, think of the organization and its market(s) as the equivalent of Newton's "mass." His second law states, "Acceleration is produced when a force acts on a mass. The greater the mass [of the object being accelerated] the greater the amount of force needed [to accelerate it]." The more extensive your strategy, the more of your organization you'll have to engage to supply it momentum; in other words, the greater the mass, the greater the energy required. In this case the motivating energy comes from the amount of resources that you commit to the initiative. The faster you want the initiative to accelerate, the broader the deployment or involvement required to motivate it. Resistance to its acceleration comes from the number of initiatives you attempt to complete during the business cycle and the rate of improvement in business performance that's desired (or required). In this description, Newton's formula of F=MA is modified from Force = Mass X Acceleration into Force = Manpower X Actions. Your realized "rate of acceleration" for the implementation will depend on how skillfully you balance the breadth of deployment with number of actions initiated.

Balancing energy and mass into the rate of improvement (force) that's required by the strategic plan is the responsibility of the senior leadership team. Implementing too few actions (too little force) is as detrimental to success as burying the organization in initiatives (too much resistance). Each induces a different type of failure, and this is where Newton's third law applies: "For every action, there is an equal and opposite reaction." The diversion of too much manpower toward the strategic tasks will remove equal attention from your day-to-day activities, and induce tactical failure. Commit too few resources, and the strategy will be stillborn. It's always a balancing act when shifting resources from daily business activities toward strategic initiatives but the performance of your tactical imperatives (the ones that pay the bills) must be sustained with a balance and timing that assures the success of both efforts.

When it comes to results, initiating excessive initiatives (mass) absent sufficient manpower (energy) will cause critical objectives to fall late or incomplete. Conversely, significant dedication of resources (energy) applied to an inadequate number of initiatives (mass) will leave parts of the organization outside of the objective set and unable to participate, formulating a "drag" on the force component. The strategy then fails to gain sufficient momentum on its own. Both examples can leave people unmotivated, or even drive them to be dissatisfied and negative, presenting the leadership with even bigger engagement challenges.

Preparing for Change

To ready the organization for its change initiative, first try to estimate the degree of change a new strategy will impose. Confirm your list of both the operational impacts, as well as the physical ones, in order to determine the level of preparation (planning, training, communication) that's required to prepare employees to adopt the specific changes.

Performing an impact assessment similar to the example below will provide "directional," insight that need not penetrate to extreme levels of detail. These work better in the absence of hard data and because the assessment is far more observational, it merely needs to be close enough to initiate a first cut at resource deployment. Look at the project list from a couple of different perspectives. Begin by ranking the list of things that *must* change, (chapter one) as equal in priority to activities associated with running the business on a daily basis. Next, look at each of the business functions and try to estimate the workload imposed by each of the following factors: 1) maintenance of tactical performance; 2) current or pending project assignments; and 3) processes that will be changing. It's effective enough to use approximate percentages of impact for each, as shown:

Function	Tactical Load	Project Load	Process Changes	Total
Leadership	75%	10%	15%	100%
Customer Mgmt.	90%	15%	15%	120%
Product / Service Delivery	90%	30%	20%	140%
IT (Info Management)	50%	20%	30%	100%
Human Resources	70%	15%	25%	110%
Finance	80%	20%	30%	130%

The sample estimates would indicate that four areas are potentially overloaded. Realistically, given the range of error possible in the directional estimates, the high totals for product service / delivery and finance would be the only concerns to take action on. At this point, you should move to balance them through resource allocation, changes in program scope or timing, or deferral of results.

Having estimated the impact to functional workloads, the senior team is now positioned to revisit the must change list and attempt a different directional assessment for the level of mass each might represent. The only true estimate of mass is going to come from looking at how the initiatives will task each of the functions involved. Next, revisit the list of things that cannot change, introduced earlier, to look for the organizational energy required to keep those untouchables firmly in place. Do so by approximating the involvement required to sustain monitoring and metrics for each.

One of the most commonly identified "cannot change" attributes is the need for a sales team to maintain its existing "benchmark" level of service to the customer. On the provision the senior staff agrees that the actual performance value measures up as a benchmark, there would be two directional effects to the organization. First, it would suggest that no negative structural or staffing changes could be allowed to impact the sales team. Second, the team's participation in the new initiatives would be limited to avoid diluting its customer service task loads and compromise sales performance. Although customer service levels certainly have to be maintained, allowing one function to have a lower level of participation or even to opt out of cross-functional involvement in the change initiative isn't a workable situation. With that, a compromise will need to be reached.

It's important to understand the impact of the "must-change" and "can't change" lists on the initiatives and the organization (the mass). Adjusting resource requirements (energy) to achieve the desired rate of change (resistance) allows the organization to achieve the new strategy in line with the expectations of both the leadership and the organization.

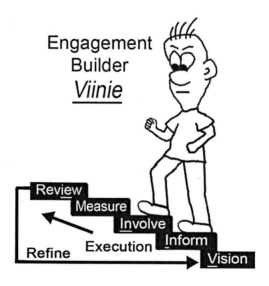

VIINIE

VIINIE is an acronym to describe a change management process that's structured so that the key elements of communication and involvement are present to help build the engagement that supports your change initiative. VIINIE's components include Vision / Inform / INvolve / Impact measurement / rEview & refine. He was created to provide a sequence of execution, as well as a refinement loop to the engagement in your cultural change initiative. This supports our emphasis in chapter 2 about the essential need to communicate the vision. In the VIINIE model, the *vision* forms the basis for the information that will be communicated to the organization.

Executing the engagement process begins with *Inform*ing the organization through a comprehensive communication process that manages content, delivery, and timing. All three elements are required for the communication process to be effective and sustainable, with the goal to disseminate the correct amount of information to each level of the organization. In doing so, it should translate the vision into content that will make associates aware of both the internal and external reasons for the changes, the approach you'll be using to address them, who will be involved or affected, and what you expect to accomplish. If you can discern a hint of R=A+D in the method, you should! Effective communication of the vision goes a long way toward forming the "roots" of your associates' endorsement and participation.

The *INvolve* step is much more comprehensive. Your choice of strategic solutions will undoubtedly impose new demands on your organization. To again use the example of a marketing function that's attempting to migrate from traditional approaches to take advantage of the acceleration in internet traffic, it's a change that could create a need for someone with specialized skills in internet marketing. The organizational skills assessment will quickly identify whether these marketing capabilities exist internally, if potential developmental candidates exist, or if an outside search is needed to hire the correct talent. Finding these gaps in the organization and establishing action plans to close them is critical to your execution of the strategy. As well, the timeline for closure must align with the strategy's performance improvement goals.

Another key component of the Involve step is a training and development plan designed to close the skills gaps that were identified as internal

Lowell J. Puls

needs. This can rapidly open opportunities for high-potential associates and accelerate your ability to leverage the organization's adoption of the strategy. Carefully selected internal development candidates can have a far greater cultural engagement value and a shorter learning curve than might the typical outside hire.

Measuring Impact

People are responsive to what you measure; or are they? While they will generally respond to what gets both measured *and* monitored, timely reviews of progress on initiative metrics help maintain the pace of activities, sustaining organizational focus and supporting the desired level of cultural engagement. Keeping associates thoroughly informed with regard to the nature of the strategy, while tasking them with activities that link directly to it, will retain their attention and their focus. Visible and active measurement as well as follow-up on the progress of implementation, will keep them engaged.

Even though we introduced metric theory in chapter 3 and promised to dive more deeply into the mechanics in chapter 8, it's adequate to mention it here as a leveraging tool – more for engagement than simply to gauge performance. Used judiciously, Impact measurements that include top-level public metrics, specific project measures, and personal targets, all facilitate engagement. Organizations that master change, succeed through relentless preparation for both the selection of initiatives and their targets, but with equal effort placed toward acclimating the organization to the strategy. Using separate measures for the engagement elements you deem critical to the execution of your strategy – things like associate development, organizational alignment, deployment of activities, etc., is a foundational step toward managing the depth of cultural change needed to achieve the strategy.

The last steps in the VIINIE process, and the final elements of preparation and managing change involve *rEview, and refine*. It's a lot to cram into only one letter of the acronym, but if I'd plugged all of them in it just wouldn't have been a usable name any longer. These steps close the "loop" of improvement because the metrics monitor execution through *Review* cycles utilizing *live* results to guide tactical adjustments through a *Refine* step, ultimately feeding back into the Vision. It's this use of active results to drive adjustments that knits engagement together and builds belief in the integrity of the process.

As the initiative progresses in step by step fashion, from communication of the vision to informing people of your expectations of them, involving them in the planning, execution, and assignment of tasks, your associates will be drawn into engaging themselves (see "pull behavior" coming up in Chapter 6). Maintain the visibility and delivery of metrics that are grounded in reality, represent visible progress, offer a public view of the strategy's status, and convey the drivers for refinements through the execution phase; and your success is assured.

Readiness to Launch

Preparation for change takes careful consideration for the magnitude of its impact on the organization, the changes in skills required, and in the involvement it demands. Taking this on without a specific *Approach*, or without understanding the intended *Results*, will leave gaps in *Deployment*. With a good situation assessment completed, engagement momentum is accelerated by keeping those who might normally be "left out in the cold," inside and informed – avoiding unnecessary conflicts between the strategy and the culture. You must continually work the plan *live* because of its ability to affect people, and anything that affects people impacts morale and culture.

Chapter 5 will give you techniques to sort through the process of prioritization in a way you might not have imagined. This brings to mind another old phase; "When you have alligators snapping at your heels, it's hard to remember that your task was to drain the swamp." If that were really the situation, both the swamp and the alligators would have to be dealt with. How to do that is coming up.

Chapter 5:
Things That Will Kill You, Things That Can Eat You, Things That Make You Happy!
Realistic priority management

1) Mastering Priorities
2) Things That
3) Things That Can
4) Things That

KILL YOU

EAT YOU

MAKE YOU HAPPY

Squarely in the midst of rolling out the newly identified strategic initiatives, our largest and most loyal customer called us to request an urgent meeting at its headquarters. We immediately interrupted our planning sessions to assemble a team of key leaders and customer representatives for the trip.

Upon arrival, we were led to a large conference room, exactly on time. The room was already full of their people, with seating locations for our contingent assigned by name. Every one of the customer representatives remained seated as we entered. Our team found their seats.

After a brief round of introductions, the company's president rose to address us. "We have been a key customer of yours for nearly half a century, and to some extent we have protected you as our sole supplier in your product category. Lately, however, your poor performance has begun to cost us business and is disrupting our achievement of our own objectives. The situation has recently degraded to a point that we're no longer able to tolerate. Our purpose for inviting you this morning is to provide a proverbial shot across your bow, and serve notice of the need for you to elevate your performance to a minimum acceptable level within the next ninety days. If you cannot, we'll be forced to transfer 25% of your business to the next most appropriate supplier, one who also happens to be your main competitor. Today, we're willing to work with you to define a mutual set of performance objectives that meet our minimum requirements, as well as to establish an expectation of rapid and significant progress on the key issues."

Even with my short association in the business, I knew we should have expected something like this. It's hard to describe the knot in my stomach as he ended his comments, but it was obviously my turn to step up and speak. Since we were keenly aware of the impact our performance problems were having on this particular customers, we had prepared well, and I felt confident of the quality of my response.

"First, let me thank you for the opportunity to be here. We realize we've given you every motive to pull the business from us. The fact you're giving us an opportunity to maintain our long-standing business relationship is a reflection of your commitment to us as your business partner and it is sincerely appreciated. I'd like to make you aware that a number of your concerns have already been identified and are being addressed in our operating plan for the coming year. Nevertheless, we are willing to return to those plans and modify them in order to give you the improvements you require in a timeframe that will support a renewed relationship." My response seemed to satisfy most of their contingent.

"We all appreciate your commitment and are confident you'll achieve the objectives we set for you today," their president said optimistically. "But I would like to get it all down in writing so that nothing is left to chance or mis-communication."

"I understand," I said. "My team and I are ready to start right now."

True to his word, the balance of the meeting was spent reaching agreement on the improvements we would be held accountable for at the end of the ninety-day period. We left the meeting troubled because we knew there was a tremendous challenge ahead of us, but equally relieved that our customer had the courtesy to give us an opportunity to improve our performance. It rarely happens that way.

Although we had just hosted this company's CEO in our strategic offsite planning sessions; he had failed to impart either the same sense of urgency or the radical timeline that had just been impressed upon us by his own president. Many of their concerns regarding improvements in delivery and quality were already accounted for in our strategy deployment plans, as were order entry and customer service enhancements. Regrettably, it was clear that our current timelines wouldn't generate enough improvement in time to meet the 90-day deadline.

We needed to adjust our priorities quickly, and although all of us wanted to sustain the positive momentum we were enjoying on our strategic deployment process, it would have to take a step back. Once we concluded the meetings and returned to our offices, we reconvened for a late evening planning session.

"Well, folks, we have three months to salvage our relationship with our biggest customer," I said. "That's no secret. The issue at hand is, are we united in making it happen?"

"I think we're all clear on what we need to accomplish," Geoff said, as heads nodded in unified agreement.

It had been a long day, but I pressed on. "Let me tell you about how I've managed to re-focus my priorities over all these years in this business. I segregate the activities into three groupings. The first group is called "Things that will kill you." Obviously these have to be addressed with the utmost urgency and focus. The roots of this customers issues fall into that category. The second priority group is "Things that will eat you." Some of the eaters have crept their way onto the list of killers, particularly in the area of quality. For now, we'll seek only to contain them until we've dealt with the killers, then return and fix them. There is a third level of priority items that I call "Things that will make you happy," and they are in essence the project that we will set aside for at least 90 days."

"Sounds kind of like triage," said Geoff. "when I was in the Army, you were taught to handle the gaping wounds first, the serious but not life-threatening problems second, and let the cuts and bruises and scrapes pretty much handle themselves."

"Not a bad comparison," I agreed. "Okay, the floor is open. Talk to me."

"You know, I obviously couldn't say this in the meeting, but our customer is at least partially responsible for some of our delivery problems by the way they forecast and order," Janet said softly.

"Anyone have an idea on how to respond to Janet's comment?" I asked.

"We might have to adjust our planning and production to be capable of responding to any possible needs that may arise within their typical ordering timeframe. Even the unreasonable ones!," Geoff ventured.

"Precisely," I concurred. "I believe it's our Response-Ability!"

"Why don't we make this customer a priority for all of our resources – in other words – give them preferential treatment?" Mary suggested.

I shook my head. "No, my experience tells me that picking between our customers is the wrong thing to do. In a way, we would just be pushing the problem around and risking the alienation of other customers – the old carpet and bubble adage. We need to review our systems and adjust them to get some real short-term improvement, even if it's at a cost."

To leverage our new-found sense of urgency and turn it into fast but accurate action, we elevated the project plan for customer performance improvement from the strategy sessions, pulled the team together for an emergency session, and began to revise both the problem statement and project list. We applied the timing as demanded by the customer's president to the problem statement and project lists we'd already developed. From there, we commandeered a war room and convened a series of daily meetings in order to drive our containment actions for immediate implementation and monitoring. Placing the process on the war room wall and making it *visual* also had value to the other associates, as they could sense the common urgency, see the actions, effort, and timing requirements. These would set a format for containment of the customer issues, while at the same time provide data for use in establishing continuous improvement actions.

"Now, as to the containment activities," I started, "we need to conduct them on a cost is no object basis to make sure they fully contain the quality and order management issues. That way, they can drive an immediate impact to the customer. But well also need a feedback loop."

"Why don't we establish a cross-functional containment team to work through the details?" Janet said. "I think Geoff and I should co-champion it."

"Make sure that in the process of doing so, that you record absolutely any and all related data, so that we can assign a second team to find root causes of those issues," John suggested. "Then they can also take the lead on implementing the improvements."

"That's a great approach!" I could barely contain my enthusiasm. "It should help minimize the duration and cost of the containment activities. Please ensure that each member of the team is given an appropriate level of reprioritization. Don't let any other killers drop through, even though this gets top priority."

You're in a swamp surrounded by alligators. The water is too shallow to swim in and the bottom is too mucky to run. How do you decide what to do? The first thing you have to consider is that the alligators could kill you, and if getting away isn't a real option, you'll either need to outsmart them or find a way to defend yourself. With every move they make, you're fighting panic, and frantically looking for options. You're also getting increasingly worried about the ones you can't see, and your senses are alive to feel if you can detect one coming from underneath the water's surface. Time isn't your enemy, for every motion might contribute to their aggression Each move you make must be calculated to improve your defensive position. You look around to see if they're moving toward you, but they appear to be stationary, checking to see if you are a threat, or possibly dinner.

Mastering priorities

In 1943, psychologist Abraham Maslow developed his view of man's "Hierarchy of Needs." His classification of essential human needs began with survival and evolved through safety, security, social needs, and self-esteem before ending with self-actualization. I studied Maslow a bit while in college, and his theories had a definite impact. As my experience in the business world accumulated, a simpler hierarchy of my own began to develop. It was influenced by Maslow's thinking, yet entailed a simplified but dramatized approach to categorizing and establishing priorities for the organizations under my direction. My personal hierarchy developed around the prioritization of business improvement initiatives and had only three levels for determining what to work on and in what order.

Establishing and sustaining priorities is a far different endeavor at the organizational level than it is at a personal one, primarily due to the number of individuals affected. Viewed personally, you have considerable control over the factors by which you rank and address your priorities, with the exception of any external interference (which I'm excluding but not discounting). At the organizational level, however, the priorities you establish as a leader will face exponential challenges. The formula goes like this:

$$\text{Priority execution} = \frac{\text{Organizational engagement (\# people)}}{(\text{\# of people involved} \times \text{Internal initiatives}) \times \text{External Issues}}$$

Got it? The equation isn't an attempt to quantify prioritization, but rather attempts to recognize and assign magnitude to the array of influences that can dilute the attention of your associates, potentially distracting them enough to drive them to work on the wrong things. Keeping their projects correctly prioritized helps them to stay focused on the most critical objectives. The graphic nature of the descriptions used in my priority hierarchy is intended to stimulate peoples into making better choices. Let's start by describing all three more fully before we delve into how they influence each other.

"Things that Kill You" bring with them the peril of imminent negative consequences. Killers possess an ability to so disrupt the business as to disable it, and they can come from many directions: problems with large customers, poor cash flow, or product and service quality spills. They require fast and assertive action to contain their damage before you can proceed toward controlling and correcting them. One of the most perilous side effects of a killer is its ability to become all-consuming of the organization's attention and resources, placing at risk many other activities that are also on the must do list. I'll refer to these as TTKYs or "Killers" for the balance of the book.

Of a lesser priority are "Things that Eat You." These tend to have a more erosive effect, differing from the calamity-creating TTKYs. Eaters derive their name from a tendency to erode your performance over a longer period of time. They can sometimes be inconspicuous, subtle, much like a disease, making them easy to overlook or ignore. Perhaps because of their less visible consequences or their slow rate of effect, organizations frequently fail to address them, being consumed working on greater urgencies (like TTKYs). With time, things that eat you have the potential to grow into TTKYs if not contained and corrected. I'll refer to these as TTEYs or "Eaters" from here on.

KISS: Keep It Simple and Sustainable

The last group of priorities consists of "Things that Make You Happy." These belong at the top of the pyramid - the self-actualization level as it was described by Maslow. They are priorities that, while potentially important to the business, fail to address (either directly or indirectly) any TTKYs or TTEYs that might exist. Although everyone desires happiness, pursuing it at the ignorance of issues that could destroy or damage a business is simply denial. Since happiness is a requirement that necessarily must come after the avoidance of being killed or eaten, it should be worked on either after the first two are eliminated or when they are contained and resource availability allows for it. One key reason for classifying TTMYH's as such is because their performance impact is always less immediate. It's their own inability to generate enough short-term improvement to offset their consumption of resources that prevents them from ascending to the head of the priority list, even when their long-term impact might be significant.

Again, this approach to classifying priorities has always been highly effective at getting people's attention. The graphic nature of the terms is an important tool in helping employees maintain focus and sort through their own activities.

So let's look at how each type of priority affects the business on a day-to-day basis. "Things that kill you" rightfully garner most of the organization's attention and all of its sense of urgency. The immediacy of a severely negative impact creates a mandate for urgent response. Killers can be simple problems disguised by a set of complex circumstances that makes them very difficult to resolve. TTKYs usually require rapid action to contain the problem in order to protect the customer and possibly the business. Once containment is achieved, it's imperative to perform a thorough assessment of the potential root causes in order to determine and implement the right permanent corrective action.

An example corrective process is one I call the three Cs: contain, cause, correct. The technique suggests that you must quickly isolate the list of potential causes and contain them so they are either stopped at the origin or kept within the business without being released to the customer or market. With effective containment in place, you can move resources to work on corrective action using good disciplined problem-solving tools. When a solution is uncovered, deploying it as quickly as is practical will speed your recovery, benefitting both the business and its customers. The negative

67

impact of a Killer can take many forms: lost customers and market share, sudden drops in business levels and lost revenue due to quality or shipment issues, quality-related problems that have exponentially negative potential (recall or liability costs), the loss of supply for critical business needs, or even just the erosion of profitability. These are areas where a significant and sudden negative change can endanger your market or operations so severely as to be almost unrecoverable. Although a TTKY can sneak up on you, your response, dedication of resources, and drive for improvement should be sufficient to neutralize the potentially negative impact.

Killers are the most difficult of the three priority levels to work through the 3 C's. Many organizations become proficient at contain and cause but lose focus or are diverted away before implementing permanent corrective actions. These behaviors will tend to become pervasive, yielding cultures that are reactive or even jumpy and consistently fail to close out problems, spending excessive amounts of time in fire-fighting mode. One concerning side effect you should be aware of is that containing them can leave an organization so "out of breath" that it struggles to attend to its daily business.

"Things that Eat You" can be nearly as lethal as TTKYs, but their rate of attack is usually much slower, leaving their causes and the resultant damage lurking far enough below the organization's radar that they're often ignored. While being eaten can be painful, it rarely conveys the level of immediate threat that a killer has. Because of this, Eaters won't attract the attention or resources that a TTKY will. Also, because of its covert profile (analogous to cancer), it can be a much easier decision to sacrifice some short-term performance and keep the organization focused on greater urgencies, such as its TTKY's. A former boss of mine used to say, "It's more important to stop the patient on the gurney from bleeding to death than to splint his broken leg."

Beyond the fact that their slow rate of attack can lull one into thinking they are only a nuisance, one other critical TTEY characteristic helps them stay off of the priority list. Their lack of an apparent cause will often convince people who become aware of them that they will require far more resources than can be justified against their visible return on investment. The higher levels of urgency and visibility commanded by the Killers will usually shove Eaters to the back seat, and the presence of a high number of Killers might even push them off the list entirely.

KISS: Keep It Simple and Sustainable

Some examples of TTEYs include dropped calls in a call center, order cancellations, unclear service diagnostic results, manufacturing scrap, process yield problems, and field failure rates. Let's look at an example of an Eater that can quickly migrate into a Killer. You're having account servicing problems which are affecting all of your customers, but they've led to cancellations only among some of your smaller clients, and it has masked the overall impact. So far, the issue hasn't been great enough to draw attention and the revenue impact has also been negligble. As a consequence, the root causes of the problems haven't been getting much attention. Suddenly, you're called on the carpet by a major account, who notifies you of the company's intent to cancel its business due to the same problems that have been eating away at the lower-level customers. Now, the issue has become such a threat to near term business performance that it leaps into the Killer category.

Another example is out of the automotive parts business. An assumed low-level quality yield problem within a manufacturing location suddenly appears in the field as a compromise to long-term product reliability; almost instantly, the component becomes a subject of a potential recall. Again, the cost of correcting the problem in the field is exponentially greater than that of containing the problem in the factory, and the issue jumps priority levels.

Because of their elusive causes, the greater effort required to correct Eaters is driven by the difficulties posed in chasing, isolating and correcting them due to the extensive process driven solutions they might require. It's not uncommon for their causes to appear to come from multiple directions and to have a cumulative impact, necessitating extensive data gathering and analysis before it's possible to plan your corrective action. As the data accumulates, the business impact becomes clearer in terms of resource and time requirements at which point prioritization can be assigned intelligently. Resource and time intensive tools, such as Six Sigma's DMAIC (Define, Measure, Analyze, Improve, Correct) or the logical problem solving approach of PDCA (Plan, Do, Check, Act) are often used to find the root causes. Culturally, the problem with Eaters is that appearance is everything, and it allows all attention to be quickly drawn toward the Killers. A focus on only the visible impact ends up as the "iceberg syndrome", since the damage caused by TTEYs is usually greater below the waterline.

Even though "Things that Make You Happy" should always be last, they

69

wander onto the middle of the priority list for many reasons. For one, their strategic value might be subjectively inflated above that of other near-term activities. A project that is more strategic in nature generally suggests a short term impact that's below that of other higher value or more rapidly implementing tactical programs. The reality is that we are sometimes happy to divert resources from those higher value programs if the perception of value is in error. A second reason might be that they offer psychological relief from the stress of more immediate urgencies since they tend to be more positive and forward looking than the more perilous Killers and Eaters. It could be a conscious decision to jump a TTMYH ahead in priority just to seek relief from the stresses of high pressure priorities.

TTMYHs can look like a capital improvement, a social or benefits program, or even a tempting acquisition opportunity. It might even be a customer account that isn't concentric to the core business. All of these can be falsely elevated in priority, even when they don't really link to a near-term need. However you identify them, get them stopped to prevent the resource drainage that can accelerate the negative effects of killers and eaters before you're even aware of the conflict.

Because they are usually easy to identify, sorting and classifying your killers and TTMYH's on the priority list can be quickly accomplished. It's the eaters in the middle that require the most scrutiny, since they can be killers in disguise. TTMYH's can also find their way into the middle group any time they get a leadership sponsor, so it takes some discipline to keep them in their proper place. As we stated, the risk of letting happy projects into the middle lies in the diversion of critical resources away from more critical problems. Allowing a killer into the middle can be disastrous – particularly if it's pushed there by an excessive number of them. If that happens, you should employ outside resources to assist in their containment and correction as quickly as possible.

For every issue it's important to identify the key business and customer facing process breakdowns that contribute to the problem and allow them onto the priority list. From there, improvement efforts should be tied to the orderly collection, arrangement, and review of the pertinent data. The right approach can vary with the challenge, but taking a high-level view of the business with the use of process or activity mapping tools is often a great place to start. Fill them out with data to see how things are flowing

KISS: Keep It Simple and Sustainable

-- products, cash, sales closures, and engineering programs. Add any known error or defect rates to better understand what's dropping through, and integrate the results into the review to estimate the resources and time required to fix each matter. With starting points identified, establish your priorities and execute the plan – *as quickly as you can!*

So, we're at the midpoint of the book, having covered vision, response-ability, R+A+D simplicity, change management, and prioritization. All of these have attempted to address the "what" for your tasks ahead, and the remaining chapters are structured to help you sort through the "how." Since we've only covered parts of the old "who, what, where, when, why, and how" to this point, I'll assign the "who, where, when, and why" to you, because it falls within your leadership response-ability.

Chapter 6 deals with helping associates understand how to respond to a business issue once they've chosen to engage in its success. We'll talk about how to use "pull behaviors" to support a compelling engagement motive.

My Machiavelli quote for this chapter? "Entrepreneurs are simply those who understand that there is little difference between obstacle and opportunity and are able to turn both to their advantage."

Don't let the things that eat you accumulate until they kill you! Start working ahead of them now!

Chapter 6:
Creating "Behavioral Pull"
People Systems that Trigger Team Achievement

1) Pull Behaviors
2) Blame *ME*, Please!
3) Request
4) Provide
5) Guide

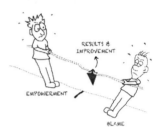

As we began to tackle the containment activities necessary to resolve our immediate customer challenges, it was becoming clear that those resolutions would have to be more than just systemic (planning, execution, and quality related) for us to improve our product and services performance. There were clearly behavioral issues that were enabling the problems to be much worse than they should have been. In order for us to move as quickly as possible, we divided the effort between the leadership team and the functional management group. While the functional teams started to work on containing the systemic issues, the leadership team began to assess the behavioral elements that were exacerbating the problem.

"In order for us to achieve these improvements quickly enough, we'll need to identify the organizational behaviors that are impairing our performance at Valhalla Corporation and then develop a plan to change them as rapidly as possible. To accomplish that, we'll first need to figure out how to do so. Does anyone have any thoughts on how to approach it? I looked back at the team to gauge their non-verbal response.

"We need to get focused on this main customer!" Dick responded

"But I'm afraid, if we focus on just the one customer, the entire effort will be viewed as a fire drill and the cultural impact won't last." Mary was right on point with her statement.

"The thoughtfulness of that comment is very much appreciated, Mary," I started. "Defaulting to my Lean Enterprise experiences, it seems they would tell us that we need to be totally focused on all of our customers. Do we think that we are?"

"Of course we aren't," Janet quipped. "Or we'd be servicing them exactly the way they want us to."

"So isn't that the first objective to establish, along with the set of behaviors that we need to change?"

"Sure it is." Dick was back in the game.

"Okay, so if the behaviors we want are for everyone in the organization to become more customer-focused, then the example behaviors will be needed from the people who are customer facing or involved in fulfillment, right?" Janet had offered me the perfect setup.

"Maybe, but if we ask for those behaviors from those who directly service the customer, and someone in a support role to them doesn't have the same level of commitment, won't it break down?" I asked them.

"So, we need everyone in the organization to be customer focused!" Karen was thinking out loud.

"Why don't we just get them into an all-employee meeting and tell them what's going on so that they understand the situation and have a chance to buy in? Geoff asked

"I don't think that will be enough," I answered. "We'll need to employ the second Lean principle and engage them in the solutions."

"How do we do that?" John (engineering) asked. "Particularly with my guys, who rarely interact with the customer and probably won't get the chance."

"We'll need to employ a Lean-based technique I've developed called "pull behaviors." I answered. "It's based on creating a compelling and personal reason for each and every associate to commit to customer satisfaction. External customer needs go without saying, but internal customers should also be identified and allowed to visibly see that their actions are connected externally – in order to ground their personal motives to the effort. As an example, if an accounting procedure is making an order late,

KISS: Keep It Simple and Sustainable

all parties involved need to engage to resolve it in the customers' time requirements – not ours."

Mary was clearly intrigued. "I'm interested in seeing how that will work."

"We need to demonstrate that there is something in it for them, and to do so we'll need to start *with* them." I responded. "We'll can begin by interviewing our customer-facing associates to see what barriers keep them from responding to the customer with the right answer or timing. We can then connect the dots backward to the support people and processes, getting the team to take decisive action to fix them so they can move at the *customers* "need for speed." Who wants to make a recommendation as to how we go about it?"

We agreed to break into two teams, with one group working on the customer facing people and another working on the support groups. The seven of us moved through the organization quickly, talking to people about their performance barriers with the primary objective of understanding, a) What were the obstructions to servicing the customer in his time frames?, b) What benefits might they personally gain from the improvements?, and c) What, in their opinion was the first step they would personally take to fix it? It didn't take long for us to learn that while the organization uniformly cared about performance at the customer, our associates felt powerless to correct the causative factors, because most of them seemed to be outside of their control. As well, many of those causes were considered to be byproducts of the design of our product and service delivery processes. It was becoming clear that the fastest way to gain traction on the issues would be to empower associates to redesign and change the processes in order to make a difference, but "how" was still the challenge.

Our first step was to communicate to them frankly and honestly about the current status with our customer and the potential impact of not achieving their mandates. We were open and honest about what had been learned through the course of the interviews and had used the information to frame a "burning platform for change." From there, we presented our management contract – a list of guaranteed outcomes from the interview sessions. Lastly, we closed with a short list of our expectations for them (mostly behavioral) and requested their participation in developing the targeted solutions. It was clear that the thoroughness of our approach and unprecedented level of open communication was harnessing their attention and building interest and excitement around the possibility of success.

Although we did take steps to place a few non-business-critical issues on hold, for the most part, we asked the team to volunteer their participation in developing the solutions. Our management-contract listed the specific areas where certain short term expectations would give way to the successful resolution of the customer issues.

With the engagement and motivational momentum building, we reconvened the functional leadership teams and set them to work on designing the process solutions.

"Behavioral pull requires each process to have a "trigger" mechanism that alerts the associate responsible for it of the need for action" I began. "These "triggers" can be tied to a specific problem occurrence or a response that covers a group of possible issues. A trigger associated with a single problem can be easier to train for, while a trigger that covers multiple issues will require a two levels of training. Once you've identified the type and number of triggers to use, your training design should equip every one of the associates running the process to provide the correct solution. Can someone think of an example?"

"I would think the best triggers would either be visual or audible." Jerry, one of our engineers offered. "In fact, different triggers might be used for a response provided by the cell operators, as opposed to one that requires outside assistance. I can see lights and sirens for certain machine functions that the operator would correct, but also see a different set of lights to alert maintenance, engineering, management, or the quality team."

"Great!" I jumped in. He was definitely getting the gist of the process. Let's set you loose to make lists of all of the triggers you want for each process, and the training needed to implement them."

The team proceeded to develop the visual indicators to help people to recognize the problems. Any items that were related to flow and subject to constantly changing status (e.g.; product quality, inventory levels, flow of services, and orders backing up), were each assigned triggers that were created by the users themselves.

The approach had multiple benefits. First, empowering the associates to design the triggers themselves helped to build ownership in their use. Second, the type of response was also conveyed visually. Finally, the triggers

KISS: Keep It Simple and Sustainable

were structured in such a way that the parameters or "boundaries" for decision-making were also visually supported, closing the loop of requesting, providing, and guiding the response. With the system complete, we were confident that any abnormal conditions would receive the correct response in time to prevent a failure at the customer.

One by one, we implemented similar systems for each of the business processes that the teams had identified. Any abnormalities in those processes began to quickly receive the appropriate attention, were corrected more quickly, and became symbolic of our success with the approach.

A core tool in Lean Enterprise is the use of "pull" systems to manage how processes respond to changes in demand. On a Lean factory floor, "pull" is most simply defined as - make nothing until an external customer or downstream process (an internal customer) requests or takes one ('pulls' it). The opposite of 'pull' systems are 'push' systems, where an upstream process produces what, how much, and when it wants – regardless of what a customer actually demands.

Applied correctly in a sequential process environment, "pull" systems make for remarkably efficient processing, with minimal work in process, little wasted effort, higher quality, and production output that is closely linked to real demand or consumption. This is the exact opposite of a push system, which consumes precious materials and resources (labor, capacity, and time) in a way that's out of sequence with customer needs, making timely supply of true market demand difficult or even impossible. Push systems always generate waste!

"Behavioral Pull" applies the principles of Lean pull systems to the performance of your work teams, but even its simplest form is more complex than just making and using something. Behavioral pull mimics the essence of the Lean Pull "need-respond" sequence by elevating the organization's ability to recognize the differences between normal and abnormal process performance and then provide the appropriate response.

The first step in establishing behavioral pull is to clearly define normal versus abnormal for a given process or activity, because it isn't always obvious. Next, it's necessary to identify the organizational behaviors required to maintain a normally performing business or manufacturing process as well as the responses necessary to correct specific abnormalities. Finally, processes are redesigned to include built-in visual and instructional cues

77

Lowell J. Puls

that will request the needed maintenance behaviors as well as the corrective problem responses. In order for behavioral pull to work in the same way Lean Pull does, a request- response sequence should be broken down into separate components. These include the following: *request* (a need trigger), *provide* (a trained, disciplined response to the defined needs) and *guide* (visual instructions that support the correct response).

Blame ME, Please!

Blame-oriented cultures represent a strong undercurrent to successful empowerment. The 'reaction' of assigning blame to a situation invokes fault and immediately overrides the more correct assumption that most associates are trying to do both a good job and the "right thing" in the course of performing their work. When something goes wrong, the "who" involved can be an important data point in the pursuit of the root cause, but it's commonly feared because of the negativity of being associated with "what went wrong." In real life, most problems result from a blend of circumstances that unfortunately place the individual at the center. Improper training, poor preparation, and preventable or unanticipated process problems, are at fault more often than are carelessness or malicious intent. Blame cultures will seek the person too quickly, and *might* pursue the other causes afterward, or maybe they won't!

To begin moving your culture away from blame orientation, two things must happen. First, you should take assertive steps to separate blame from accountability, because it is possible to hold people accountable without invoking blame. Second it's important to delay the assignment of responsibility for a problem until you've looked thoroughly at all of the underlying facts – especially when they aren't immediately apparent. Begin constructing the *"what,"* (as in what happened) in painstaking detail by starting at the point that the problem surfaces and use a technique such as "5 Whys." Even when some of the causative factors point to an individual, avoid jumping to the *"who"* until after all of the "what" data suggests that you should. Once an individual's performance is actually determined to be a component of the causes, the corrective action should focus on process construction, training, and error proofing before performance! These are viable techniques to eliminate the potential for human error without defaulting to the person (see chapter 7 – Jidoka and error proofing). You'll have to give people breaks in favor of a relentless pursuit of root cause, in order to begin winning over the culture and turning off the "blame-throwers."

Logical problem-solving techniques support blame elimination by discouraging what I call "jumping to contusions"; hasty/incorrect presumptions of cause that end in pain and difficulty, and will slow the evaluation process through misdirection. These techniques can further subdue defensiveness through disciplined root cause work, and with it go the barriers and apprehension that might impede the corrective actions, making the rest infinitely easier. Using work teams to pursue and eliminate the specific cause(s) builds "forward ownership" in the solution, as well as a team-based environment that engenders behavioral support for the corrective process changes.

Using team-based logical problem-solving techniques to clearly identify root cause, and supporting them with structured process-improvement tools to implement sound solutions, will assist you in cleansing blame from your culture. Keeping the organization focused on physical remedies as opposed to individual responsibilities also produces more sustainable solutions. To understand this better requires a key concept in Lean "Error Proofing": if a (process) defect has a statistical possibility of occurring, then it will occur. I've always referred to this as "pollution in the stream," because although the problem may be so small that you may never be able to see it, the statistical potential for its existence means it *is or will be* there at any point in time. The sustainability of your solution depends on the extent to which it can eliminate every possibility of the problem occurring.

Recognizing Abnormalities

An associates' ability to provide a response will depend on his/her understanding of the difference between normal process operation and abnormal performance, as well as the factors that create each condition. Complex processes will have a variety of contributing factors which, when performing abnormally or below expectations, can require different responses. Each issue must be defined, and resolved with a specifically trained response to sustain normal process performance.

Your processes should be redesigned so that the existence of an abnormality is visually triggered, and associate training is put into place to enable them to recognize the trigger and respond to it with a containment or

Lowell J. Puls

corrective action. As noted above, normal process operating limits should be visually identifiable as well as the response trigger points and the corrective actions necessary to return performance to target.

Request

So let's look at how behavioral pull employs a "trigger" to request a specific behavior or "response" to any of a group of identified process abnormalities. The process should be designed in such a way that a visual cue is activated to expose the problem as quickly as it occurs, hence the name "trigger." The trigger design will specifically ask someone responsible for the process to intervene with specific action that is supported through training. Using an example from a payables process; an abnormality occurring from an accumulation of too many pending accounts payable, initiates a response request. The "trigger" designed by the team is a simple physical height gauge next to the payables inbox which visually indicates the number of invoices waiting in cue. Once the stack reaches a height indicated by a color coded mark on the gauge, the associates who are trained to process payables, respond by shifting to the process – adding capacity and reducing the backlog.

Another example is in a manufacturing process where production activity is controlled by visual locations on the floor that instruct associates to start or stop the process based the presence or amount of inventory in them. If the locations are filled, no product is to be produced, but if they are emptied and the markings are visible, their color codes (red, yellow, green) visually instruct the process operators as to how much product to make and how urgently to respond. Yet another example might be a customer service process (you've seen this in large retail stores) where the number of people in line (possibly identified by markings on the floor or the wall) will trigger customer service associates to respond by moving from other activities to the service process that is backing up.

The common element of these behavioral requests is their use of visual or audible cues to specifically request those associates assigned to the process to respond as they are trained. A trigger must clearly expose the problem and explain the need for a response, with training supporting the type and quality of response that's associated with the trigger (*guiding*) and will resolve the problem.

Provide

With the abnormalities identified, the corrective actions defined, and the behavioral responses targeted, training is required to enable associates to recognize and respond correctly to a specific problem request. The extent of the training required is dependent on the process type, but it's critical to point out that both training and response-abilities should be layered across the organizational levels to "engage" everyone associated with the process. Whether the person is charged with operating the process and recognizing the existence of the issue, or supervising the area with responsibility to support and reinforce the response, it's the combination of training, the layered responsibilities, and the visual controls that makes the response to a process issue "reflexive."

By "reflexive," I'm referring to a response that becomes "subconsciously automatic" from the respondent. You can't accomplish this kind of response in a single event, but you will build it with a combination of training and standardization that's sufficiently clear, thorough, and so well reinforced that the correct response occurs without external intervention. The more complex the process and the response required, the more frequently it should be reinforced. My own personal example of "reflexive" came from a period when I was wrestling for a Division I university. Our coach would drill us on certain techniques until we were literally sick of them. At the time it seemed to be more punitive than beneficial, until my return to the sport ten years later to coach my own sons. Even after so much time away, my ability to react to live situations with the correct technical responses was surprising. It had taken me that long to learn and appreciate that he'd trained us so thoroughly that our physical responses had moved beyond a conscious decision to a subconscious response that I call "reflexive.

Guide

Once your process controls are established, they should be integrated into the visual system along with their respective abnormality responses. User training is necessary to be capable of associating the visual cues with the proper actions. Making the process control elements visual is a fundamental of principle of Lean, and by default, of behavioral pull as well. The visuals support and strengthen the value of the training, helping to semiautomate a repeatable response when a trigger is activated. All of these contribute to your ability to sustain normal process performance.

With triggers that are designed to visually communicate as much information as is possible about the existence and nature of the issue, other process visuals will help to request the appropriate response. Again, the visuals are intended to solicit and support associate intervention for all aspects of process performance. This includes normal functions, flow controls for quality and rate of output, and intervention actions to counteract abnormalities, involvement, the timing or urgency of the response, and more.

Let's look at an example of a visually controlled customer service process;

	Visual Controls		Layers of "Pulled Behavior"		
	Normal	Abnormal	Associate	Supervision	Management
Service Flow	Customers in line equal 5 or fewer		Continue operation	Normal monitoring	Random monitoring – verbal
		More than 5 customers in line	Change lane light to yellow	Increased monitoring Support associate – redeployment; open a new lane	Increase attendant staff if problem persists
Problem Response	Service lane light is on – white in color		Continue operation	Monitor as appropriate	Random monitoring – verbal cuing
		Lane light is flashing Yellow	Available associates open a new lane	No associate available: Self- staff a lane or elevate w/ intercom	Self-staff a lane or call in additional Associates

In the above example, visual cues are lighted (lane lights) or add an audible trigger (intercom). Each level of the organization is trained to provide a different response based on the type of cue.

Now, let's look at an example of a visually supported manufacturing process;

KISS: Keep It Simple and Sustainable

	Visual Controls		Layers of "Pulled Behavior"		
	Normal	Abnormal	Associate	Supervision	Management
Scheduling	Material present in Staging zones – raw (In)	Empty in or out staging zones	Produce to schedule Resolve and /or elevate	Normal monitoring Resolve when elevated	Random Monitoring Inquire when aware, involve as needed
Production/ Flow	Process is operating (green light)	Operation is ceased (red light)	Produce to schedule Resolve problem or elevate	Normal Monitoring Respond to visual cue & resolve	Random monitoring Inquire when aware, resolve as needed
	Material present in staging zones - finished (out)	Staging area is empty	Resolve or seek help	Ensure solution or help	Inquire / Involve as needed
Quality	Good product staged		Continue operation	Visual monitoring	Visual monitoring
		Products red-tagged on- hold	Seek assistance (red light)	Respond as cue requests	Be aware & involve as required

In this example, the cues include lighting and color-coded staging areas. The staging locations are designed to accommodate a limited amount of moving material with the requirement that, should they be filled, the process must stop. The general design of each of the cues and triggers is standardized across the organization, however, the cues vary by process and the type of problem and response requests can be specific.

The management task associated with behavioral pull is to engage associate input and participation in the design of the visual controls, the responses, and also the training. This helps them to commit more fully to the program and provide intervention as required by the rules of response. On a day-to-day basis, management supports continuous improvement through timely resource allocation, priority management, and more consistent expectations. This is essential, because if the expectations of the process become variable or you allow priorities to change excessively, the consistency of results will be lost. Like the rest of the organization, managers should be trained thoroughly in the use and meaning of the visual controls, their ongoing role in day-to-day reinforcement of them, and how to establish expectations consistent with the desired response. From a support perspective, the set of expectations must also require full participation in training from all levels of associates.

Summary

So, the visuals engage by exposing the problem, requesting intervention, and guiding the response. The training enables associates to reflexively provide the correct support to the trigger's request. Management commitment is demonstrated through support for the investment in training, the establishment of consistent expectations, and the monitoring of results. The cultural impact of all of this is "pull behavior." Converting it back to R=A+D; the correctly trained response is the desired result that comes from your approach to training, with responsibilities deployed to the appropriate associates for corrective action. Implemented thoroughly and with measured patience, behavioral pull will improve response and engagement, paving the way to more rapid resolution of problems, higher performance and sustainable improvement.

But, what does it take to break it all down? The common mistakes made when trying to enlist engagement include: 1) underestimating resistance by failing to understand its origins within the key involvement group; 2) trying to go it alone, or involving too few associates in the process design – flying solo prevents you from earning the team's buy-in and reduces organizational momentum; 3) trying to solve the problem with only a technical solution, which rarely works when there is heavy influence by the human element; and 4) failing to celebrate success at the same pace as the achievements. Excessive celebrating can create the appearance of unwelcome hype, whereas too little suggests a lack of appreciation.

"The beatings will continue until morale improves." Having worked for a couple of guys who managed with that approach, it seemed odd that they never understood the way it drives people into their shells, preventing them

KISS: Keep It Simple and Sustainable

from reaching out to help one another, severely inhibiting creative input, and keeping them from stretching beyond a guaranteed result. These were organizations that always underperformed. Lean has taught us that engagement is a key to achieving otherworldly levels of success and satisfaction.

So far, we've looked at simple methods for maximizing engagement by steering both leadership and associate behaviors. Chapter 7 will look at how Lean tools and thinking can be used to accelerate our progress even more.

Chapter 7:
Takt-ical Management
Using the Lean Tools to Build Improvement

1) The Lean Commandments
2) Achieving Flow
3) Avoiding Suboptimization
4) Managing with Cycles of Response
5) Using the Lean Tools
6) Implementing Lean

"Folks, I'd like you to meet Jorge Ribero. He is a Lean Enterprise expert whom I've asked to come in and help us gain some traction on improving the order fulfillment of our operations."

"Greetings, I will appreciate it if everyone here can arrange to meet with me at some time today," he began in his thick Brazilian accent. "My evaluation will take only a couple of days. However, I need your opinions about the issues in your operation to help me find the correct solutions to suggest to your directors."

Several of the customer issues we had highlighted were rooted to deeper problems that became apparent as we worked on applying visual and behavioral pull solutions. The team had struggled with Lean Enterprise implementation for years, getting inconsistent results that were rarely sustained. Now, as they began to dig more deeply into several of the most pressing problems, it was clear that an inadequate commitment to Lean was a key contributor. Since Lean had already been identified as one of our critical strategies, it seemed wise to enlist Jorge's expertise to help us determine a course of action toward putting our Lean effort to work. He and I were familiar with each other, and his skilled observations would make a tremendous difference.

He spent the better part of the next three days interviewing the leadership team, the functional managers, and key contributors, both to gain insight into some of our business practices and issues, as well as to see if we had any internal candidates capable of leading our Lean effort. Jorge meticulously combed through our manufacturing processes, asking questions, jotting notes, and offering a few observations. He was looking specifically for those areas where we could quickly improve the flow of our processes in order to have an impact on all customers, not just our unhappy one. On the fourth day, we came together to hear his report.

"Some things you are doing very well, some not so well," he said, as he paced the conference room. On occasion his accent was difficult to understand, but the meaning behind everything he said was very clear. His delivery was as concise as if he had rehearsed it to maximize its efficiency.

"I have reviewed your production processes and spoken with your key people. Mostly, I see a good capability to perform Lean, without significant problems with attitude. The quality of workflow is poor, and there would appear to be something like 15-20% of improved productivity available. People are working hard, but not in a Lean manner. This is what you will seek to change, and your business will perform better."

"I can't buy into a 20% productivity gain." Geoff seemed almost offended by the idea.

"This you can achieve and more. Your productivity will be gained not from what you see, but from what you do not see," Jorge added. "Let me explain!"

With that, he moved to a whiteboard and began to draw a couple of pictures. "Your flow loss is coming from several problems. One is lack of material. Not lack of sufficient material, but one of readily available material."

"Please, explain." Geoff was suddenly getting interested.

"By ready material, I am referring to the fact that you are committing material to products in a way that is not in the same sequence as the market is demanding. When the market demands something, you already have it committed to something else and you consider it no longer available. Thus, yes, you have sufficient material, but it is not ready."

"But we've processed it for another order," said Geoff.

"Not an order." Jorge paused. "You are doing so for a variety of reasons. You want to use your machine capacity and your manpower resources efficiently." He went on. "When you use up material to be efficient, but you do so on products that are not needed just to utilize capacity, it is waste. This is so because you are not actually making things to sell today. When you make products not being asked for, it is inefficient. It is worse than that when you make products that are not in demand and you use valuable material that cannot be replaced until your suppliers can respond. Just because you want to keep workers busy, you forfeit the ability to fill a different order for another eight weeks. Even so, the workers will come back tomorrow with possibly nothing to do. Both the workers' efforts and the materials used are misguided."

Geoff sat very quietly for a moment before responding in an almost questioning tone. "But, with only sixteen weeks to deliver a product that takes twenty weeks to build, something has to be ready before an order is accepted."

"I agree, but you have options." Jorge smiled at Geoff. "Let me show you them." He explained his recommended solutions in some detail, providing directional estimates for how much improvement opportunity might be available and his thoughts on how long it would take to achieve it. "Before I leave, there is one more critical observation I would like to make," Jorge added. "Your paper processes do not flow at the same speed of the work in the rest of the business and, therefore, they hold it back. This, too, you must correct. If you manufacture at takt, your papers must flow at takt as well.

From there, he answered a few questions from the team and made a closing comment or two. As he began to pack for departure, I was compelled to approach him. "Jorge, I appreciate all of your comments, but I'm still not certain of exactly where to start."

"Flow from the customer order point for each product," he responded abruptly as if I hadn't been listening. Nothing more was offered in response as he headed to his rental car.

My staff and I met immediately afterward. "What do you suppose he meant by *that*?" I asked them.

Geoff spoke first. "I think he means we should start after assembly, which is where the point of order fulfillment is established."

Karen, the plant controller, spoke next: "Well, we learned that we should start at the customer end of the process, so maybe what he was saying was that we should start where the product gets committed and then flow from there."

After considerable discussion, we agreed that his message was for us to achieve flow from the point a product was committed to a customer or order, which actually varied by product. Karen's observation that we should start at the end of the process and work to achieve flow moving backward into the facility was spot on. In contrast to a typical constraint busting approach, flowing from the customer at its rate of demand would produce immediate improvements in response, stabilize output at a more predictable level, and provide a sequenced game plan for improvement. While chasing bottlenecks might provide for more dramatic early improvements, it would also have the team scampering around the processes in search of the next bottleneck. Since the net results are always dependent on the next constraint, the game plan would constantly be morphing itself. Worse yet, instead of turning the focus on the fulfillment of specific customer needs, we would instead be focused on the chase for process performance.

During the next several months, we employed some external guidance along with supplemental training to re-energize our Lean transformation. We tackled the entire fulfillment process, beginning with the highest impact products for specific customers. Our problems weren't with the highest volume items, but rather some middle and low volume niche products with more erratic order patterns. By working a parallel path to flow manufacturing, and at the same time streamlining the paper processes, we elevated our performance for the customer quickly enough for us to exit its performance plan after only the first round of improvements. In addition, our other products produced in those and similar value streams benefitted from the same improvements, helping us increase sales by nearly 15%, and increasing our profitability on those products by eight times while our four walls productivity improvements surpassed Jorge's prediction, rising 30%. This first really cohesive Lean implementation was a runaway success, a clear victory for everyone that also saved us from a possible business-altering change with a key customer.

Most often associated with manufacturing, Lean principles apply to the improvement of any and every business activity conceivable. Techniques that originated around streamlining factory workflow have been transferred

KISS: Keep It Simple and Sustainable

successfully into restaurants, retail stores, office environments, hospitals, and even construction and refinery sites. Lean's relentless focus on the elimination of waste through the creation of an efficient flow of value added activities allows it to apply with immense impact to almost every business activity.

Because so many books have been dedicated to the topic, I won't delve any more deeply into it than is necessary to simplify the way you apply it to your change initiative and clarify the impact it can have on sustainment. I don't consider myself a Lean expert, as that honor is reserved for those who think about it a lot more than I do. My experience in Lean spans nearly thirty years at the tool level and another twenty developing my practitioner skills, where I've achieved tremendous success and learned to appreciate its power. We'll cover Lean at this point in the book for two reasons; first, the simplicity inherent to Lean solutions is fundamental to the charter of this book. Second, as a career Lean practitioner, I'd like to try to simplify and promote its magic, while leaving the technical and tool details to others. For that reason, let's take a look at Lean's synergies to K.I.S.S.!

Lean Commandments

As a highly experienced implementer of Lean, my belief is that there are just a few core "commandments" that cannot be ignored. Let's try to understand what they are, how they can help improve performance, and how failing to attend to them can impede progress.

1.   **Start with the Customer:** The primary objective of your Lean effort is to become adaptive enough to respond to any level of market demand in the timeframe your customer or market wants it. Consider your customer to be the "creator" of demand, and define him as anyone from a business patron, to an associate sitting next door who is the downstream user of your product or service (from material goods to data). What's imperative to understand about the customer relationship beyond the type of products or services he buys, is the rate in which he consumes them. That rate of market demand is referred to as *takt* in Lean speak, and it's the fundamental building block for the design of all of your delivery processes for goods or services. Being certain you're structured to handle market

91

and customer needs at the rate they are required is a key success factor. We'll expand on *takt* shortly.

2. **Engage your people:** The second commandment is nearly as important as the first. The people in your organization possess the skills and capability to work through its challenges and sustain or improve its performance. But, because it requires their complete commitment, you must "manage to engage" your associates as a key success factor for leadership. Begin with the set of business leading indicators you have already defined (markets, strategy, takt, economy), and provide your associates with the information and direction that will motivate them to focus their efforts on improvement. Engagement that immerses them in the success of the business also gives them a personal stake. Lean is a heavily people-dependent process where one of the most common failure modes is one of not engaging your team.

   A sizeable part of this book has been devoted to the topic of engagement because it's so critical for its ability to involve the associates who are closest to the problems in their resolution. The Lean tools will equip them with the skills to correct issues more rapidly, and engagement harnesses the collective power of their intelligence and experience to achieve accelerated solutions. Their buy-in congeals into the glue of sustainability, but it will all dissolve under inconsistent management actions that break down the developing disciplines before they can take root.

3. **Find your *takt* and use it** - A German word meaning rhythm – takt represents the pace of demand in the marketplace. Understanding demand in your own market (and its patterns if any) is the most logical starting point to determine how to satisfy your customer's needs. The takt calculation converts demand into a time factor or, the amount of time you have to provide a unit of product or service to keep pace with the rate of demand and the formula for takt-time is a simple one:

$$Takt\text{-}time = \text{demand} \text{ / } \text{time available}$$

KISS: Keep It Simple and Sustainable

Where 'demand' is the known market requirement for your product or service; and 'time available' is a conversion of the resources you have to fulfill your share of market demand, into time. You may choose to level, average, or share-factor demand in order to develop the most usable picture of any market. 'Time available' can be based upon a person, a machine or machines, a function, or a department, even an entire office complex or factory that supplies the specific product or service. A simple iteration of *takt* is the requirement to ship one unit during every minute of operation. I want to emphasize that although the usual interpretation of *takt* refers to pieces or units of output, we shouldn't limit ourselves to that definition. Substituting activities (for instance, customer service opportunities per hour, patient arrivals; etc.) in lieu of units of output are examples of how to apply it to your own processes.

Although the calculation of *takt* can become challenging when you're trying to combine different markets, products, and/or service offerings, the conversion to *takt time*, can almost get contentious. It becomes very easy to confuse *takt-time* with process cycle time and it's important to keep a laser focus on the difference. Variable demand and the impact it has on *takt-times* can be further complicated by diverse fulfillment processes that must consider equally variable resources; equipment, disparate parts of the organization, even sources of supply. Different segments of the organization might sometimes compete for control of specific elements of the fulfillment processes, making it even less predictable. Fractioned response systems might have multiple or cumulative response cycles and result in sub-optimized systems. It takes a homogenized version of *takt/takt-time* to determine the response cycles for the entire organization – as well as the proper resources to ensure it remains Lean throughout. Remember that the basic purpose of *takt time* is to convert demand into resource time so that units of output can be measured against the resource inputs. Every one of your fulfillment and support processes must be structured to perform at the rate of overall *takt (within takt time)*, and it takes discipline of leadership to focus your organization on the simplest yet most effective method.

4. **Seek "Flow"** – The objective of the Lean tools is to achieve a smooth flow of the activities necessary to deliver your product or services at the rate your customers demand them (*takt*). That's a mouthful, and it's far harder to achieve than it sounds. If every segment of your fulfillment processes performs and is supported at *takt*, the response cycles should connect to yield continuous flow. I'll emphasize the word "supported" to stress that fulfillment processes which are able to flow at *takt* are quite often constrained by the administrative processes that surround them. That's why it's essential to build the same response timing into them that will enable the rest of the organization to respond to demand, otherwise it will result in rework activities throughout the organization, disconnecting flow, adding cost, and dulling response. To accomplish this, you'll need to use *takt* as a basis to reconstruct your fulfillment processes with a focus on process simplification, alignment to *takt-time*, and interconnections that will establish and sustain an operating rhythm that supports its attainment.

Achieving Flow

Thruput – With *takt* defined you'll next have to determine your ability to deliver those product / services within *takt-time*. I'll call this thruput, which is a common, but not universal term used to define the current response capability of your fulfillment processes. Thruput-time is the conversion of your delivery system to a time metric, and different than *takt-time*, it focuses on the total aggregate time required for a product or service to take shape through your processes before it can end up in the hands of a satisfied customer (ex; three days of thruput time, with 1 unit completing every minute at takt time). The *takt-time* in a service business could require that you process one person through a line every five minutes, even though the total time your customers invest in you from the moment they call or enter your waiting line to the point you complete their requests for services might be a half hour. In manufacturing thruput can represent the total time required to bring a product from raw material to shipment. Thruput-time has two components: 1) Processing or Cycle Time; and 2) Movement or Waiting time.

As you begin your Lean effort, it's necessary for you to understand

KISS: Keep It Simple and Sustainable

thruput-time and its breakdown in order for you to determine your ability to improve response. As you do, you'll generally find movement/waiting time to be much larger than a process or cycle time. Though both can be improved, focusing first on non-process time can usually provide greater gains for less cost and effort. You can find thruput time by adding cycle (process) times and inventory cues (measured in days of usage), using a stopwatch to measure it directly (like a service), or simply by attaching a tag to a representative product and asking each person who touches it to check it in and out. Exercises of this type are necessary to define and address your Response-ability.

Processing time is usually viewed as value added (not waste), whereas movement and waiting will take the form of either non-value added time (waste) or necessary non-value added (unavoidable waste). The Lean objective is to build systems to guarantee continuous flow from the start of the process to its finish. Improving that flow involves separating waste from the value added activities in each process step and eliminating it. From there you can more easily streamline the value added work, enabling greater efficiency and faster thruput, while preserving the critical customer deliverables.

Pitch - Alternatively, there is yet another component to the thruput/response equation that can be useful as an aid in planning for fulfillment. It is referred to as pitch, and it's used differently by various Lean practitioners, but I prefer to use it to describe an increment of flow that is established by either the business or the marketplace.

A real-life example of pitch might be a dozen eggs, which is equal to a pitch of 12. I'll offer no research into why eggs are predominantly packaged in dozens; chickens certainly don't lay them 12 at a time! But the dozen egg carton is the predominant method of offering them to the market. Although pitch can be a derivative of the method of transportation, with eggs it's more likely evolved either by habit or tradition. Because a dozen eggs is so entrenched in the way the market buys them, the effort to change market pitch could be more costly or difficult than any gains from doing so.

Another, more definable example of pitch might be socks, which are used in pairs and sold in pairs, but not as likely produced in pairs unless it's in batches that are divisible by two. Since anything sold in pairs as both right and left hand versions has to be matched, gloves and shoes are most obvious, pitch can be used as a tool to simplify flow through the manufacturing, packaging and distribution processes. Pitch is an aspect of demand/flow that is important, because ignoring it can result in waste, but leveraging it has the power to reduce costs.

Avoiding Sub-optimization

Practicing Lean necessitates an awareness of the concept of sub-optimization. This occurs when you increase performance within a segment or subset of a process, but do not improve the output of the production or service system, thereby failing to obtain an overall benefit. Both factory and business processes suffer from sub-optimization, and it's usually self-inflicted. I've seen many organizations invest capital in additional machines and office equipment to improve the thruput of a part of their process only to ignore the main problem or relocate it elsewhere and gain far less than the value of the investment. Avoiding sub-optimization requires discipline, preparation, and a thorough understanding of your entire process or system. It also necessitates a clear picture of the demand that flows through the process and alignment with the resources available to fulfill that demand.

One example of sub-optimization can be found at a retail store where the staffing is perfectly resourced to assist customers with their selection or to replenish purchased products on the shelves, but an inadequate number of store associates with cashier skills leave customers standing in line and frustrated, or even abandoning their purchases. The reverse is also possible, where cashiers are standing at their stations waiting for customers, but there is no one to assist with product selection.

When I stress that your administrative processes should be aligned to have the same response capability as your "physical" or conversion processes, the most common example usually occurs in the area of order entry. In my own career, I've seen countless examples where the activities around order receipt and processing consume far more time than the processes required for creating and delivering the product or service. Another example comes from business planning, where a plan constructed to fulfill a specific level of market demand becomes disheveled by fluctuating market activity, stressing the organization's internal and external resources. The impact creates imbalances that move back and forth from materials supply, to the fulfillment processes, and back to demand planning, depending on how effectively *takt* has been used to manage the alignment of these components.

Solutions that are deployed in an un-level or untimely fashion to only one or two of these inputs result in sub-optimized performance. These imbalances always have both cost and business-level (revenue) repercussions.

The concept of sub-optimization requires that you always maintain a "system level view" of your improvement efforts. System level implies that you should identify flow and quality improvements that have the greatest, most direct impact on your ability to improve response time. Projects that don't have the appropriate amount of system impact should be delayed until they do, or eliminated entirely.

Managing with cycles of response

Having covered the Lean Commandments, it's time to introduce another key success factor for the deployment of Lean. The creation of a continuous flow of process activities requires all systems to move in harmony with one another, and this timing must be replicated throughout all of the business functions. I'll refer to these as "cycles of response" for the way they can synchronize all business activities to align with the *takt-time* for the business. In a manufacturing environment, successful response to changes in customer demand is the cumulative product of aligning four primary response cycles: a) one for your supply chain (goods or services), b) one for your internal fulfillment processes, c) a management cycle to align the administrative processes, and d) a customer cycle that represents the time between their request for goods or services and the expected delivery of them. Aligning each of these to a version of *takt-time* is an aspect that is too often overlooked in many Lean implementations.

Let's start by identifying the response cycles that cannot be changed directly – the ones we must align to rather than placing them in alignment to ourselves. The most important of these is the time the market allows for delivery of products or services in response to a request, without losing customers or impeding market growth. Often called "Lead Time," this should usually be determined by your competitive environment. "Market-defined" means you have to align to it, and you can't change it without redefining the requirement by developing a product or process that's so compelling people will wait for it, or by developing response capabilities that exceed those of the competition, thereby reducing lead time for the market and gaining a competitive advantage. Another way to gain advantage is by subdividing your market response offerings, such as faster response time for higher volume or competitive offerings and slower response on low volume or custom ones. Dependability of performance is a huge response advantage – the more dependable you are, the more customers can rely on you – and they will! How innovative you are in building your market response-model can become a game-changer that leads you to success, but without improvements like those, you'll have to align or lose share.

The second cycle that also can't be changed (although we have more influence in this case) is the time it takes for suppliers of the products and services used in the business to respond to *our* needs. Good management practices, such as the placement of strategic inventory, dual capacity models, or supply agreements that cover supply timing or different business levels such as volume and mix changes, can facilitate response during times of market change. Still, it's constructive to understand that you don't have absolute control over it.

These two cycles establish the bookends for the design of your own response systems. The open challenge is to for you construct your internal systems, both demand fulfillment as well as management processes, so that they are as robust as possible and can adapt quickly to changes in the market, or disruptions in the two external cycles. Great system designs will blend internal management controls, processing requirements, capacity and investment (people, equipment, inventory, and facilities) into a usable formula that will enable you to adjust *takt-time* to changes in market *takt* fluidly, and inside of that market's lead time. If you can't adjust capacity to match changes in demand, you will either turn business away or spend yourself into a less profitable competitive position.

Active adjustments to market changes are more readily done in environments that manage using highly competent forecasts for demand, supply, and capacity consumption. In a Lean environment, you define your market *takt,* then structure your delivery processes – for services or hard goods - with the appropriate facilities, equipment, and staffing to enable you to meet that *takt.* That accomplished, your ability to respond will be limited to your least adaptable process.

There are some key differences between office processes, and service or manufacturing ones. For office and (sometimes for) service processes, *takt* can be difficult to determine. Both can lack a defined scope of work and consistent process thruput times that prevent you from determining time available. Because office processes typically aren't constructed and measured with the same scrutiny as are manufacturing or some other service process, aligning them to *takt* can take considerable effort. Also, because the staffing decisions for these processes tend to be based on "positions" that don't actually have measured work to enable their conversion into *takt-time*, staffing levels become "fixed" and less able to adjust tasks, increase performance, or monitor response to changes in demand unless someone with sufficient experiential knowledge intervenes. Many businesses are constrained by the performance of their administrative processes; causing uneven demand response that is rarely mitigated. The effects snowball into the performance of the fulfillment processes, creating a diversionary effect that makes the real problem even less visible.

Matching the performance cycles of administrative processes to the fulfillment processes <u>will</u> improve the performance of your business. If an abnormality in a fulfillment process requires a corrective response (an example might be an emergency purchase), then your approval and processing steps to procure the item must be able to respond in the same cycle time as the need. By understanding and converting the administrative workflow cycle times to match takt time, they can be incorporated into your process design to ensure timely response. You can then establish a response capability for your business processes that's in sync with the market needs.

In building business-wide cycles of response, you must identify the key processes across the entire business that can affect its performance. From there, you'll need to prioritize them in the order needed to confirm the basic process functionality, align their cycle times, and determine the level of total resources needed to sustain a timely response. When we get to metrics in Chapter 8, these cycles of response will be extended into your review and feedback activities, which should facilitate your ability to generate timely corrective action and implement process improvement.

Using the Lean Tools

No matter what Lean expert you talk to, he or she will usually define the set of Lean Tools differently. I suppose it's an approach that provides an appearance of individual proprietary knowledge with respect to Lean methods. My hands-on use of the tools in manufacturing's trenches has led me to create a functional arrangement of the tools, grouped under what always seemed to be the key themes in Lean: continuous flow of activities, standardized work, problem response and resolution, and continuous improvement.

In the 1999 Harvard Business Review article, Decoding the DNA of the Toyota Production System,[1] the authors highlighted four key attributes of Lean which should serve as the pillars of your own response processes. They are paraphrased below:

1. All work is highly specified for content, sequence, timing, and outcome.
2. Every customer to supplier connection is direct.
3. The pathway for delivery of each product or service is simple.
4. Improvement methods are logical, mentored, and performed at the lowest level.

Each of these themes requires a separate focus and an integrated effort in order to achieve true Lean results and commensurate success. The tool groups contain individual Lean tools that interrelate and implement in a complementary fashion. I'm not planning to go all the way into the detailed tools in this book, because again, there are others who have thoroughly blazed that trail. My purpose is to provide insight about the main themes, why they exist, how they relate, and how to simplify their use and avoid complexity.

Flow: R=A+D

Lean Tool Group	Result	Approach (Tools)	Deployment (Process)
Flow	Connect customers and suppliers	Value stream mapping / *takt*	5 Value Stream design elements: *takt*, activity flow, inventory, process design, customer order point
	Simplify the product / service path	Continuous flow	
	Deliver to the customer	Pull Systems	

---

1 [1]Decoding the DNA of the Toyota Production System C1999 by Harvard Business Review - S.Spear and H.K.Bowen

KISS: Keep It Simple and Sustainable

Flow Tools

This group includes Value Stream Mapping (VSM), continuous flow processing and pull systems. They are coupled for their collective ability to identify and facilitate the flow of activities. Used as the primary planning tool for a Lean effort, value stream or process mapping begins with *takt* and collects all of the process data required to enable you to understand and expose the inner workings of a value stream. Populated with the correct data, the map serves as a structured planning tool by which you can establish improvement priorities.

A thoroughly prepared map helps to identify opportunities for flow improvements, and is also helpful in selecting projects that use the other Lean tools to facilitate linkage into a continuous flow of activities. Fulfilling customer needs through the use of pull system buffers and triggers, allows the system to respond fluidly to changes in demand and assists you in establishing predictable performance. Lean approaches further minimize investment in assets, inventory, and manpower.

The flow tools work in unison to focus your efforts on process sequences and waste elimination, shortening processing time and reducing your cost for products and services without losing business due to forsaken demand.

Standard Work: R=A+D

Lean Tool Group	Result	Approach (Tools)	Deployment (Process)
Standard Work	Work content and sequence is highly specified	5s / Standard work	1) Define *takt* 2) Major step 3) Key points 4) Reason why
	The value stream is continuously improving	Standard work	
	Cost objectives are obtained	Total Productive Maintenance (TPM)	

Standardized Work

One of Lean's most powerful and broadly applicable concepts are standardized work. Although we've already touched upon it on several occasions, it's difficult to adequately underscore its importance. The essence of standard work is simplicity, with its greatest value being in its contribution of repeatability, discipline, and predictable performance.

How is it simple? First, for the work in question, standard work asks only, "What is the major task to be completed?" In other words, what segment of the product and service delivery process am I working on here? The "Major Step" describes the primary objective of the process work sequence. It might represent a single assembly operation, processing a bill for payment, or even taking a patient's blood pressure and pulse as a function of a pre-checkup routine. Next, standard work asks, "What are the Key Points?" Again focusing on the specific process step, these are items to be monitored or watched by the person(s) executing the process. Key points can include maintaining critical quality or product features - assuring incoming quality for the next process down line, or they might include the act of completing a checklist for the client in question.

The final element in standard work involves the question, "What's the Reason Why?" Providing associates with meaning for their task steps and process verifications goes a long way toward ensuring they will consistently perform it in line with the instructions. While it may sound prescriptive, the intent of the major step is to identify clearly the work elements to be completed, in line with the key points, to insure a consistent result, always reinforced by the reasons why. Together they collaboratively insure that associates don't introduce randomness to the process that might jeopardize the desired result. In the construction of standard work, each question should be answered simply but completely, with the three elements forming the "standard" for the "work" to be completed.

The Lean tools that make up the Standardized Work group were selected for their ability to contribute to a more complete result. They include 5s (workplace cleanliness and organization), Standard Work (detailed process work scope), Total Productive Maintenance (operator sustained equipment conditions), and Setup Reduction (streamlining process changeovers). Elements of each of these tools should be combined as applicable into a master "Standard Work" document for a given process. Although certain elements of standard work will originate from various areas of the business (examples include; engineering specifications and customer quality requirements), the best approach to developing it will be written and completed by the associates who actually perform the work. Not only will the instructions be more accurate and user-friendly, but the associate's involvement in the process also will motivate them to pursue the most efficient processing options, keep their work areas in order, and monitor / maintain the performance of equipment they are assigned to.

Jidoka: R=A+D

Lean Tool Group	Result	Approach (Tools)	Deployment (Process)
Jidoka	Improvement methods are logical, mentored, and at the lowest level	Visual Management	Establish the value stream visually – use the other tools to contain, correct, improve, and sustain
	The value Stream is self-correcting	Autonomation	
	Quality is provided in the eyes of the customer	Six-sigma	

Jidoka - Problem Resolution Tools

"Jidoka" is a word that is sometimes used to refer to a group of tools and a system that helps associates identify the causes of abnormal performance, then to provide the appropriate responses for problems that will prevent or correct them in the future. In combination, the tools I group under Jidoka provide associates with the skills to self-monitor and diagnose their processes, empowering them to improve them using documented Standard Work as their baseline. This is a very powerful group of Lean tools that are too often overlooked or undervalued because of the difficulty in implementing them. Jidoka is valuable for its ability to create the momentum for continuous improvement through engagement, significantly improving sustainability.

Three lean tools comprise Jidoka: Error Proofing (the use of process based and logical problem solving tools to eliminate the ability to create a defective product or service), Visual Management (visual tools and systems that can immediately communicate the status of the workplace), and Autonomation (the analysis of the interaction between operators and the process based equipment to maximize the effectiveness of both).

Used in concert with "Behavioral Pull" techniques, Jidoka tools can fundamentally transform your cultures ability to respond to problems, correct or elevate a request for assistance, sustain the solutions, and enable consistent customer response. Thorough deployment of Jidoka can result in an environment that self-corrects itself to become progressively more responsive to minor deviations in operational performance – thereby driving sustainment.

Continuous Improvement: R=A+D

Lean Tool Group	Result	Approach (Tools)	Deployment)
Continuous Improvement	Prioritize employee safety and Wellness as #1	Value stream Mapping	Combined use of the tools to train, engage, and build a continuous improvement mindset
	Evolve to a Lean culture	Associate engagement, Feedback and Improve Loop	

Continuous Improvement

The last of my thematic grouping of Lean tools covers the development of behaviors and skills that will sustain your gains, and drive additional improvements. Setting the stage for continuous improvement involves building a consistent group of management behaviors that we'll simplistically refer to as "management commitment." This commitment consists of an implementation plan for comprehensive employee training and involvement which will develop associates Lean skills while also establishing critical energy for achieving their engagement. Included are a set of metrics which monitor and guide your program, are explainable, and are understood as applicable to all levels of the organization. Finally, the concept of continuous improvement implies that there is also a methodology for turning the newly attained benchmarks and the lessons learned in achieving them, into an evolutionary process that creates a "feedback loop" to accelerate improvement even more. These steps serve as the "cement" that will sustain and support carry-forward gains.

Implementing Lean

Not every Lean tool will be required in all environments, but there is a dependent relationship among many of them, and a well-executed Lean program will use at least one or more tools and techniques under each theme. Said differently, use the tools that will enable all activities to flow in as standardized a process as possible, with a clean, visual, responsive environment that engages associates in such a way that they diagnose and correct problems so rapidly that it continuously improves itself.

So, how can you make something as simple as this so difficult to implement? There are many ways to fail with Lean. You can overuse or underuse the tools, complicate the transformation with restrictions or unreasonable goals, even slow implementation by interjecting activities out of priority. Management must adjust its own behaviors, taking care not to impose expectations or constraints that will diffuse engagement, while fully supporting the implementation and empowering associates through a paced deployment of the necessary skills. The rate of implementation should insure that all of the dependent tools are implemented in support of one another. Even when results are impressive and momentum builds, impatience cannot be allowed to radically alter a plan that is working properly at other levels. Finally, attempting to move too quickly (outpacing the rate of learning with a compressed implementation sequence) will cause you to leave people behind, impairing active review and adjustment based upon the lessons learned.

It became clear to me early in my own Lean journey that it is incredibly counterintuitive, and each year that fact becomes more apparent. Once you learn to cast aside performance metrics achieved with sub-optimized value streams in favor of the more cohesive performance of an optimized *flowing* value stream, the benefits will be far greater. A number of years ago, I allowed an improvement team to re-flow a machining cell improperly, just because they needed learn more about the actual improvement methods themselves. While the lesson had a negative cost impact in the short term, its engagement value was a real bargain. Once they discovered their solution wouldn't achieve the goal, the team quickly reconvened and corrected the problem, eventually driving more than 25% of the labor out of the cell and reducing the overall processing time by about 30%, well above the cost of their "lesson." As an added benefit, future improvement efforts that involved

Lowell J. Puls

members of the same team were even more beneficial. The lesson for my staff and I was that imposing restrictions on their ability to fail would truly have been suboptimization.

Providing a product or service to market is easier, more predictable, and more efficient when you use the correct Lean tools with the right plan. Attempting to do it without enough tools can yield sub-optimized business performance.

"Measure only what you can act upon, and no more. When you measure, make sure that the meaning of the metric is understood by the entire organization." Jorge's words still make complete sense, years later.

Developing a set of metrics that will consistently direct your business in good times and bad, and that can be absorbed by the organization is an essential management skill. One of the potential causes of a failed Lean program is the tendency to focus on poorly defined, too many, or the wrong metrics, clinging to them by tradition rather than function. Chapter eight will help to simplify the metrics challenge.

Chapter 8:
The ABC's of SQCD+T
Simplified Metrics that Self-Improve

1) The Motive to Measure
2) Metric Types
3) Selecting Your Metrics

Our efforts to streamline our paper processes and improve customer service performance unearthed some disconnects in the definition and management of several key metrics as well as a number of lower level ones. The gaps had contributed to making our problems at the customer worse than what we were actually seeing internally. With this discovery fresh in our minds, we revisited our business objectives to clarify exactly what we should measure and how we should be measuring, before determining how we would need to deploy the changes, monitor them, and communicate them effectively.

"Well, after our deep dive into our issues at Valhalla Corporation, it's become clear to me that we're monitoring so many metrics in this business that we've lost focus on measuring what matters," I started. "Further, we aren't always taking action on what we do measure, which is the first indicator that it may not need to be measured."

"Can you give us an example?" asked Karen.

"For one, we measure factory performance from about eight different angles, and the results actually contradict themselves in a couple of cases. Of the eight factory metrics we're looking at, only two are essential to the health of the business. Let's review as a team the metrics we're tracking in the business today."

Lowell J. Puls

I slid a pair of packets across the table to each of my staff. The top one was a list of every current metric and their trends. Underneath the cover list was a thin packet of six sheets with only headings. "Please take the list of current metrics, and open them to the second page titled 'Safety/Wellness.' Let's see if we can identify the key measurement for our business in this category."

"Well, it's obviously factory safety," Geoff said.

"Is it?" I challenged. "Is factory safety alone really inclusive enough for the entire organization?"

"It does cover the greater part of the employee population," Mary said. "But maybe, to your question, it's a potential feeder metric to a broader wellness goal."

"I think we need to decide that as a team, based upon our best knowledge of the business." I was again coaching. "As always, let's make sure we take both an external and an internal view for each metric."

"How do we view safety externally?" John sounded puzzled.

"Are our products safe enough? Is there a sub-metric for safety that supports an objective to have the safest products on the market?" I asked. "What about our public image? Are there things we can do as a major local employer to make our community safer?"

Using their new-found options associated with the external view of our metrics, the team began to brainstorm ways to revise the metrics structure.

Once they had agreed to a new safety metric (triangulated between people, product and community) we wrote a one-page definition to clearly define its measurement logic. From there, we identified the pareto (next level) metrics for each subcategory that would feed the key metric and then defined staff ownership, reporting responsibilities, and the impact each department would be expected to have in the achievement of the metric target. With Safety completed we moved the next key metric.

"We did a great job getting our Safety metric to look externally and more fully encompass the business challenges we're facing. From here we need to turn all of our key metrics to look outward first, just as we did with Safety. The next two are customer quality and customer delivery or responsiveness. When I say they need to look outward, I'm again referring to finding an

KISS: Keep It Simple and Sustainable

external basis for the goal. They should be defined and measured from the customer or stakeholder's perspective."

"So, you mean the objectives set for us by the corporation?" Janet asked.

"Not entirely. The objectives set by the corporation are derived from an incremental improvement target over the prior year's baseline performance. When I say look outward, I want us to triangulate the objective from what our customer expects, what we know about our competitors, and what we think our current capability is. If we do it correctly, it may lead us to a goal that will give us breakthrough performance."

"What if we really don't know all of that information?" Dick was thinking from a highly data-driven perspective.

"Then we lean toward what we know about our customers' objectives for us."

We repeated the process used for safety by reviewing the key metric definitions individually for their ability to fulfill the measurement intent. There were few problems with the quality metric, as the top level external measure appeared to have a straightforward definition and execution that was representative of the correct intent. When we dug into the internal pareto level metrics however, we found they didn't necessarily address what the customer was seeing, and we took steps to align them.

When we got to product/service delivery, it was apparent that we'd been measuring it entirely against our own internal criteria rather than the customer's. This had played a large part in our failure to understand the severity of our performance issues with them. A customer by customer review revealed that the main driver was that each was imposing different delivery criteria and evaluating us against their specific requirements, making a homogenous key metric difficult to construct. After some considerable discussion we were finally able to modify a collective metric that could gauge our compliance to their needs instead measuring performance against our own capabilities. We continued this process over the balance of the day as we placed definitions around customer quality, customer product/service delivery, and even cost performance (two metrics: one for product cost and another for operational cost performance).

"Now let's look at time," I instructed. "When I say time, I'm thinking of our own market responsiveness."

Lowell J. Puls

"But I thought delivery was responsiveness," Dick said in a questioning tone.

"Nope, in this case we need to define a time-based objective for every organizational function that aligns with the market response time requirement. Some examples: In one organization, the financial team set a goal to be able to close the books in four days – it actually achieved it in two! In an automotive business, the development teams could regenerate product every four to six years, but needed three. Since we as a manufacturing organization must perform to "Takt," or the rhythm of demand in the marketplace, each of our administrative and support functions should have a time-based objective that aligns with takt-time at the top level and then deploys to a pareto-level metric within each function. The end goal is to comprehensively manage time in such a way that we make ourselves the leader in market response"

After extensive discussion, we agreed to a business responsiveness metric that encompassed everything from receipt of order to delivery of product right to the customer site. Each function had a contributing role to play in the improvement process, but most importantly everyone in the organization would be engaged in some way to achieve the goal of best-in-class response to the customer.

"These key metrics will be used to drive our business. We'll communicate them to all associates, and train everyone in their meaning as well as how they will be measured. Please, realize that these aren't all of the things we'll need to measure in the business, but they are all we'll report publicly. When a metric slips off track we'll drop down one level at a time until we find the pareto metrics that feed it. We'll then elevate our attention to that metric until it returns to normal and the performance problem disappears. Communication of these metrics will extend organization wide to enable everyone to clearly understand the meanings and, that we as leaders understand our problems and what has to be worked on to resolve them. Our deployment of resources will actively insure that problem resolution occurs within the needed timing."

I closed by clarifying the imperative behind the metrics program. "As the senior leadership team, our job is to help deploy the metrics across the entire organization by participating in workforce reviews and asking individual associates whether they understand a given metric well enough to explain how they can personally impact its performance."

* * *

During an interview for a very senior operations position with a company, I asked the hiring executive what he measured in his business. He proudly handed me a roughly 100 page metrics packet. Flipping through it quickly, I cautiously asked how many of those metrics he and his leadership team were truly able to take action on. When he responded, "All of them," it was clear to me that this job was of no interest to me. I'd already learned the hard way, that no organization could properly focus on or work-to-ground, that many metrics.

There are a great many organizations that over-measure their businesses. To use a common manufacturing example, they will measure headcount, labor efficiency, labor input, and direct staff versus indirect staff, indirect labor efficiency, productivity and salaried personnel. Unfortunately, the only meaningful measurement is how many people are required to provide the portfolio of services or products offered at the pace of market demand. Although they are actively measuring every individual in the building from four different angles, the common failure to perform the due diligence to understand exactly how the presence of each individual or role contributes to the value add to the business is the missing element. In Lean, we refer to this as making wasteful activities more efficient; in the end they are still waste.

A smaller number of companies tend to tie their evaluation of business performance to only one or two metrics. Whether the metric is financial or customer based, a singular focus on one bell-weather metric ignores the fact that the dynamics in any business system will limit the perspective of any single metric to a narrow, one-dimensional view. Sound metrics systems use multiple metrics with defined interrelationships to provide a three-dimensional view of performance.

Keeping your key metrics to a critical few doesn't mean you don't measure other things; it just means that all of the things you do measure aren't positioned on your dashboard. The most important reason we limit the number of key metrics to a critical few is to promote broad organizational focus. Properly chosen, four to six key metrics reside at the top, and are fed from below by pareto metrics to drive the performance of the business. The objective is to make your progress easier for everyone to understand and sign up to.

Why Measure

Let's return to R=A+D and chapter three for just a moment. At the time it was advised that you should first ask, "What result do I want" from a new strategy or initiative?" Of course, once you answer that question, it will quickly lead you to determining the approach to take in order to achieve that result, and then who you will deploy it to. The act of defining the targeted result is a big step in setting the framework for what to measure through the course of your strategy.

Very often, a new strategy alone won't create a need for new metrics, but it will very likely establish different expectations of the existing ones in order to fully support the new business objectives. The changes may necessitate modifications to their definitions, as it isn't unusual for traditional metrics to evolve over the life of a business in response to its day to day needs. While they seem to align with true needs of the business, they are often modified by corrective actions imposed to address temporary business issues. Established metrics should perform the following functions: satisfy your stakeholder and market requirements, be meaningful to everyone in the organization, have concise definitions which are clearly communicated, and be broadly and understood deployed. If they don't meet those requirements, it's time to push reset.

Structured as such, your metrics will have the best chance of being universally understood by all stakeholders, a requirement to drive the behaviors that support the business objectives, which is the real reason you measure, anyway.

Metric Types

If you asked 100 businesses to list their entire set of metrics, you might be astounded by both the size of list and the variation it contains. Some of those metrics will tend to be similar among competitive business types whether they are financial businesses, marketing businesses, distribution operations, real estate sales offices, or manufacturing entities; each has a different group of unique and critical metrics, with indigenous definitions.

KISS: Keep It Simple and Sustainable

Yet, again to reprise elements of chapter three, they should broadly include the following:

1. Safety or Wellness - Employee-focused
2. Quality - Customer-focused (external customer "touching" first, internal process-focused at the second level).
3. Delivery - Customer focused (delivery of products and services to customer needs measured against market-defined timing).
4. Cost – externally focused (cost of delivering products and services against market benchmarks)
5. Cost – internally focused (cost of operating performance)
6. Time (a time-based response metric that is market supported and applied enterprise-wide).

All of the above metrics are recommended as top-level indices - or the measurements that are placed in view of the entire organization for the purposes of communicating progress and soliciting action-based behaviors. It's normal and necessary to break these down into pareto metrics, which drop down one level at a time and "feed" data to the performance of the key metrics. These lower level metrics provide a pathway to assist in determining the root causes of performance issues and the selection of corrective actions.

For the most part, your key metrics are results-based, or rear-facing. Again, fed by pareto-level metrics they serve both as diagnostics for problem resolution as benchmarks to steer performance. The larger metrics groups break down into three categories, with the key separator being time - the amount of time it takes to review them and the time in which corrective action is expected to be initiated. Let's look at each of the three groups.

Results metrics

The time factor that segregates a results metric from the others is one that guarantees they are always be reviewed after the fact, having aggregated over a period of time – one week, one month, one quarter, or even one or more years. The time basis can vary for analysis purposes, but the results can't be affected, because the performance window is closed. Let's take the example of sales in any industry where daily, weekly, or monthly sales may be too volatile or "lumpy" to be analyzed. By extending the measurement

Lowell J. Puls

period, you can smooth over the lumps, or even some kinds of seasonality – differences that might make them helpful in managing forward or more proactively. Results metrics are necessary to ascertain progress in order to adjust your tactical direction, and to steer your strategies. It's their longer, steadier point of view, and accuracy borne of hindsight that makes them easier to communicate effectively within the greater organization. To be clear, the greatest value of a result metric is in what it can teach you – just like an historical artifact.

Managing Metrics

Identical to results metrics, the factors that separate managing metrics include both evaluation time as well as the action/effect timing. With managing metrics, the review timing is usually shorter – weeks, days, even hours, with rapid commensurate action required in order to effect the end result. To explain more fully we'll return to the sales example. In an industry where the final product might be perishable, daily or even hourly sales trends might be necessary for the management team to understand when to initiate promotions, or change the amount of inventory they bring in. Management metrics have a direct push-pull characteristic and they require tight alignment between the review timing and the initiating of corrective actions for them to function effectively.

Predictive Metrics

It's important to keep in mind that rearward facing results are useful primarily for identifying and guiding strategic and corrective actions, and management metrics can be used to steer direction and mitigate problems. While all metrics are ultimately results, there are a precious few that can predict the future by the way are trending, or by how they contribute to those trends. You can expose their usefulness after investing some time to learn how their relationships work and what review timing is best. Because they have predictive power as opposed to providing after-the-fact results; the earlier you identify and respond to these metrics with management efforts, the more these "leading, indicators" can guide you to improvement. Not all predictive measures are obvious, or even directly correlated; for example, a leading indicator to a safety metric may come from tracking the number of "near misses" or coaching discussions logged, whereas a leading indicator

KISS: Keep It Simple and Sustainable

to sales may be the timing or pattern of responses (hits) to your internet marketing campaign.

Predictive metrics are the most powerful of all for their ability to initiate steering actions which can impact future results. The most commonly used predictive metric is sales forecasting. It often (but unfortunately not always) drives marketing actions, materials purchases, human resource commitments, and other activities. Another option for sales might be internet inquiries. There are other examples; for employee wellness, your predictive options might be health care cost trends, safety incidents, or even trends for employee absenteeism. Predictive metrics require more discovery work to find, and the data often has to be "mined" for long periods of time before useful corollaries can be fully discovered. In the end, they are simply so valuable that every key business measure should have a predictive metric leading it.

Financial Metrics

Always a member of the key metric group, financial measurements have been left as a separate topic, primarily because not all organizations treat them as "public" measures. As an example, although sharing critical financial information to the associate group may be preferred in some businesses, there are many valid reasons to limit distribution – such as insider stock sales for a public company. We've included only the basic financial measurements in our table in order to acknowledge their importance, but will exclude them from any at-length analysis or discussion. That said, the simplification methods we have reviewed to this point as well as some additional ones coming up, all apply to financial metrics as well.

Metric Response Cycles

Chapter seven introduced the concept of cycles of response as they apply to the structure and execution of your business processes. The concept also applies to your metric reviews because if the review timing is kept in sync with your priority checks, your ability to successfully steer the business with corrective or preventive adjustments is improved. I'll again pick on sales for the subsequent examples because nearly every business relies on it as a key metric, making it most helpful to explain the concept. The term "metric response cycles" implies that the measurements and the response actions are inter-dependent. Since every metric data point has a limited useful life, the

115

Lowell J. Puls

timing of data collection, review, and corrective action is critical. Most of the data occurs environmentally and is beyond your control; markets, people, conditions, and other things change constantly and can either affect the metric, its baseline, or even its definition. Equally, the ability of corrective actions to alter the causes of a performance problem also changes over time. Metric response cycles require that you identify the timing requirement associated with each metric, and tightly coordinate your reviews to enable timely corrective or steering actions.

Selecting Metrics

Finalizing your list of business metrics can require a bit of trial and error. So many different measurements can be collected and reviewed, as driven by a specific type of business, that I won't even attempt a list. To get started, follow the high-level business themes to ensure that the interests of customers, stakeholders, and employees are supported. From there follow the common themes to determine what's commonly important, and therefore what to measure. Of course there are checks and balances to follow which include;

1. Maintain the top-level $SQDC^2+T$ themes! With the exception of Safety, it's critical that one metric doesn't receive preferential treatment over another, so identifying a key measurement along each theme adds balance to your measurement and review process, while keeping associates from all functions engaged in their attainment.
2. Align your metric baselines with the organization's key objectives and their initiatives. Each metric target should align with the achievement of a key objective, or you probably shouldn't be using it.
3. Triangulate to improve effectiveness (more about this shortly). Three points of reference help to legitimize the metric target.
4. Maintain the correct levels for communication, review, and initiation of actions tied to your metrics. In other words, keep them stratified by the level of the organization that will act on them. Top-level metrics will be reviewed and deployed organization-wide. First-level pareto metrics will be reviewed at the senior staff level, within a function, or co-owned by more than one function; next-level metrics can be reviewed at a functional or department level;

KISS: Keep It Simple and Sustainable

others might go all the way down to a work center. It's critical to understand that mixing the review and communication of multiple pareto levels just confuses the organization and causes associates to lose focus. The lower the level of the metric, the tighter should be the review and response group.

Triangulation

In ancient seafaring days, there were no GPS devices available to help determine your position on the open seas – in fact, if you go back far enough even the simple compass didn't exist. In order to find their way across the vast expanses of ocean, navigators used a technique called triangulation. This involved the determination of direction and relative location by viewing the ships position as it was related to two other points – typically astronomical. Since astronomy had become a well-developed science in some cultures, the sun and stars provided ample information to set a general direction. Today, GPS technology utilizes a similar technique called "trilateration" to describe their method of determining the user's position by correlating his position using the signals from the three closest GPS satellites. It's a much more accurate version of the same technique, and it applies to your change initiative as well.

My use of triangulation has evolved from practice, over the course of many years running larger organizations. Today, it's a rare occasion where I can't go into a business and reduce what's being measured, allowing people to "dial-in" on the most critical performance indicators. In each case, their performance improves and the organization becomes more successful, not less. Why do you suppose it works that way? Part of it lies in the ability to get the associates focused on what makes a difference in their performance rather than all of the ways they are being measured. Another important aspect that often goes un-noticed, is that the independent functions in a large business generally have the latitude to establish their own pareto-level metrics and include them in their functional objectives. These functional metrics can frequently fall out of alignment with the other organization-wide measures, and their timing doesn't always match with the key metrics in the business. The result is metric overlap, silo-based performance results, and misleading "cause-effect" relationships. Having too many metrics that aren't uniformly connected to the business objectives reduces focus and performance.

The technique used to manage metrics by triangulation is similar, but with a slightly different objective than the one used in ancient navigation. In this case, if two tracked metrics can provide a relative readout of another, then the need to measure the third isn't required unless there is a specific reason. It's a valid approach to separating the "must" metrics from greater group of those which "can" be measured. Comparing all of the metrics in the selection set, as well as moving up and down through the pareto levels can help you to understand the way some metrics influence others and which ones are critical to measure. Let's look at a couple of manufacturing and distribution examples. 1) If on-time delivery for your business is good, but delivery from your suppliers is poor, then chances are good that you're holding too much purchased inventory. Does it mean you do or don't have to measure purchased inventory specifically? You don't absolutely have to, but it *is* a choice you can make as indicated by its performance and the relationship it has with your primary inventory metrics. You will always absolutely measure inventory and purchased inventory is certainly a pareto level option, but attentive supply management will improve your purchased inventory and you can avoid breaking it out separately unless you find other reasons to do so. 2) Although labor efficiency and productivity are also common metrics, these too can be eliminated by monitoring labor input (manpower applied) and unit production output (takt attainment). If labor input is correct, and unit output is to takt, productivity is automatic. Labor efficiency on the other hand, is a sub-component of productivity that is so easily manipulated it's nearly worthless as a metric. Aligning the correct inputs and achieving the expected outputs can usually eliminate peripheral measurements.

Triangulation also helps to eliminate cause-effect metric relationships. These occur when one metric directly or indirectly affects one or more others. Measuring cause-effect metrics can be frustrating since actions taken on one affect the results of the others, a phenomenon that can confuse root-cause and corrective action efforts. Typically these will occur when attempting to take measurement snapshots of indices that change while moving through a sequence of which inventory is a common example.

The use of triangulation can enable a reduction in the overall metrics set and allow you to do a better job of closing in on those measures that actually drive the business performance. Whereas every business has key metrics that are essential to their daily function and are always measured, the pareto level metrics present opportunities where time can be saved and energy focused.

Below is an example chart from manufacturing for measurement, communication, and review of metrics performance at multiple organizational levels and across each of the metric types. The key metrics are balanced among the categories that were identified as essential and reviewed monthly with the entire organization. Those metrics slated for daily review among the leadership team should receive daily action with resources deployed to mitigate performance deviations and keep the business on track. Finally, pareto level metrics are reviewed by the cross-functional working teams on a weekly basis. Although this is primarily for example purposes, it incorporates all of the concepts we've reviewed in this chapter.

Results - Metrics Review and Communications Plan			
	Key Metric (Monthly Review)	Daily Review	Weekly Review
Key Metrics: SQCD+G	Results Scorecard	Forecast / Diagnostic	Pareto Level Metrics
Safety / Wellness	Safety Incident Rates	Near Misses	Injury type, Activity, Cost
	Absence Days (Wellness)	Preventive Actions Completed	Days lost by injury type
Quality	Customer reported Quality	Internal Process Quality	Supplier Quality
	Field Performance	Internal Scrap	Warranty Returns
Delivery	Product / Service Delivery / Fulfillment to Customer Request	Internal - Daily Fulfillment	Forecast Accuracy
		Order Response Time	Sales Trend by Product
		Order Aging	Late Supplier Deliveries
Cost	Inventory Metric	Daily Receipts	Inventory by Value/Model
	Productivity or Value-add Metric	Daily Cost Input / Relief	Staffing, Shipmts/Lbr Hr
+ Key initiative	On-Time / On-Cost launch	As Identified	Gate deliverables on time
	Sales From New Product / Service		Sales by product to target
Financial Performance Metrics	Total Sales	Daily Sales (Shipments)	As Identified by business type
		Order Intake	
	Profit	By Business Type	
	Working Capital / Cash Flow	Inventory Receipts	
Review Level:	Entire Organization (Public)	Management Team	Process Mgmt Teams / Workcells

Summary – "I just can't figure it out!" the apprentice carpenter said to the journeyman. "I've cut this board three times, and it's still too short."

Complex metrics breed mistakes, and though that may not have been the apprentice's problem, you need to ensure that your measurements and communications are clearly defined and executed. Once your key metrics

Lowell J. Puls

are identified and you've triangulated your way to the critical sub-metrics, what else do you need to do? Find and focus on your predictive metrics as quickly as possible. Then track them in parallel with your results metrics to understand their relationships.

For most companies, the concept of predictive or leading indicators might seem to be a "Make You Happy" priority, but the reality is that they can have dilutive power over the problems which may also be eroding your results. The sooner you can focus on them, the faster you can accelerate improvement and begin to make magical progress.

As we review the lessons learned, take steps to ensure that no matter what is measured, we always make them meaningful by doing the following;

1. Plan your measurements carefully.
2. Construct the metric definitions so they are honest, simple and explainable.
3. Keep the key metric list short: S (or W), Q, D, C^2 + T.
4. All metrics should have owners or champions assigned to them.
5. Always communicate and fully deploy the definition and purpose of the metric.
6. Consistently and visibly act on all of the key metrics (for correction or improvement).
7. Stratify your pareto-level metrics, manage them at the most appropriate level, and review them one-level-up *only!*
8. Define your cycles of response so that your review and actions create a process improvement loop.

Chapter nine will look at yet another aspect of building sustainable systems: auditing to sustain and engage. Spreading out or deploying the responsibility for embedding new behaviors and an expectation for results is an essential element of making them habitual. Auditing can be a positive methodology for doing just that, and we'll look at the approaches to accomplishing this next.

I'll summarize my metrics philosophy with a couple of quotes of my own; "Conducting measurement and review without taking related action is waste. Either cease measuring or act on the information." And finally "Pursuing the right metric target in the wrong sequence will induce poor behaviors and unravel an improvement plan more quickly than is imaginable."

Chapter 9:
"En-gage!"
Involve to Engage and Create Sustaining Behaviors

1) Involve
2) Engage
3) Standardize and Simplify
4) Audit
5) Reinforce

Having expended immense effort defining and deploying our strategy, initiating a Lean implementation, installing a commonly understood metrics system, and communicating our progress with the team, we had achieved a number of improvements very quickly. Now it was time to free up the leadership team and get them focused on long range improvements. It had to be done without decreasing organizational focus and destabilizing the rate of improvement, and it was clear to us that sustaining the initiatives we had already completed would require broader deployment in order for us to survive the constant rate of change.

"Welcome everyone!" I announced to a group which included all levels of management as well as hourly group leaders. "Let me show you some examples of the improvements that have been made in terms of our business performance over the past six months." I flipped six slides showing marked improvement in each key metric; safety, delivery, quality, cost-management, and responsiveness. "This entire team has been instrumental in these improvements and I'd like you to give yourselves a round of applause." Everyone stood and clapped.

"The work is never over, and today I'm asking for your help in making these improvements more permanent. Since even good changes can disrupt what we've just begun to do well, I want to make sure we continue to benefit from what we've learned."

With that, the staff took over, alternating roles as they explained the concept of "pull behaviors" to the room full of people. Geoff finished by asking them for their suggestions on how to "pull" through sustaining behaviors from the greater organization.

"We're improving at documenting our processes, but I think that finishing the job would be helpful!" Mark, one of the manufacturing engineers offered.

"Great comment!" Geoff replied. "But maintaining the accuracy of those documents can be a hassle. Does anyone have any ideas on how to do *that part* of it better?"

"We should have a change control procedure that is governed by the owner of the document!" the engineer responded.

"So the answer is for us to identify the document owners and their responsibilities?" Karen asked. "Good! How do we make sure that it happens? You know, people get busy and often don't get to the semi-repetitive daily tasks. Then it comes back to bite us later."

"Auditing works in the accounting world. Why don't we audit these new processes and documents?" asked Martha, the accounting manager.

"The auditing load can't fall on too narrow a group of people or it won't get done properly," offered another accountant.

"And a single group won't have the multiple perspectives necessary to do a proper audit." Mary spoke up. "If we're really going to utilize "pull behaviors" then at least some of the auditing will have to be done at a peer-to-peer level."

After discussing a variety of options, the group concluded that "layered" auditing had great potential to solidify our progress. In order for us to stay consistent with our overall themes of maximized engagement and open contribution; they suggested a couple of twists to the traditional auditing approach that would have significant value.

The first "twist" identified was that they didn't want to audit results. These were already covered by the metrics and the feeling was that auditing a result wasn't proactive or timely enough to avoid waste effectively. Instead, they elected to target the behaviors that would generate the desired results.

As a team, they spent a considerable amount of time identifying behaviors and metrics which were considered to be more diagnostic in nature, and capable of alerting associates to the need for corrective action.

The second "twist" was a plan to broaden the level of involvement and engagement to all associates, well beyond what might be true of a typical audit program. Getting as many people involved as possible would reduce an individual audits scope, simplifying it and making it an element of the "key points" of the processes "standard work." They correctly identified the increased pressure to train people in their expanded audit, response, and follow up duties. Although the primary motive was to create ownership through involvement, they also wanted to promote peer to peer coaching by having associates audit each other's sustaining process steps. The mission was to encourage real-time sustaining behaviors through positive reinforcement.

Consistent with our experience in the prior initiatives, the use of cross-functional associate teams to create our audit process earned tremendous credibility for the program. The teams designed the audit approaches, the training and screening programs, and the reward systems. They were then recruited to train the users, monitor the launch and fine tune their approach along the way.

As was expected, the groups took to task willingly enough and did a great job in lining out the preparatory work. The implementation started slowly, meeting with a bit of resistance, mostly due to our associates' struggles with peer-to-peer coaching. We tackled that particular challenge with positivity coaching sessions. The leadership team received conflict avoidance and performance management training, while the supervisor levels became conversant with facilitator skills. Everyone's individual objectives were altered to emphasize the need to manage and intervene. We took extra care to ensure that no one was penalized for a failure to adapt, and performance interventions were limited exclusively to instances where there was a failure to make an effort. Gradually we built momentum, and as the organizational behaviors began to match those we had targeted, the improvement gains accelerated.

* * *

Thus far we've discussed numerous aspects of involvement and engagement, and yet there are still a few left. Both require incredible diligence to become culturally ingrained, and the problem usually begins at the leadership level, because it can only take one innocent "executive decision," to make a direction change seem schizophrenic to the "student body." Absent a clearly communicated definition of the *why*, adjustments to the game plan can break down and disconnect your change efforts. Certainly the leadership in any business has the authority to change direction, but their dependence upon the organization for support during an implementation stipulates a reasonable level of continuity. Involving associates in the decision process and engaging them in the implementation serves to expand the brainpower applied to solving the problem and builds a commitment to deliver results, vastly increasing your odds of success.

Involve to Sustain

We'll start with involvement, which, when applied to the problem solving process is powerful enough to strengthen the organization's skills as well as the quality of the solution. Both are fundamental to sustainment because implementing a quality solution designed by an involved team, encourages them to take responsibility for the improvements, gaining acceptance and speeding deployment.

The best approach for keeping associates involved is to ensure that their role in the project, the behaviors you expect from them, and the leadership participation throughout the effort are *all* clearly defined. This is similar to the requirements we discussed under empowerment, where the definition of their role also defines their span of control, as well as the point where they should take action to pass a decision from one level or function to another. Leadership must be available to take timely action for the system to sustain its effectiveness.

Consistent with the timing concepts covered in "cycles of response," your improvement initiatives must move at a pace that is gently greater than the organizational momentum, creating that needed "pull" on behaviors. To reinforce an earlier point; moving too quickly leaves the organization disconnected, critical details unfinished, communications fragmented, and results lagging. Conversely, moving too slowly hampers the creation of organizational energy and momentum for your improvement efforts, causing you

KISS: Keep It Simple and Sustainable

to underachieve. Monitoring and adjusting the pace of the effort to generate constant positive inertia is pivotal to sustaining your improvements.

Engage to Sustain

The fundamental difference between involvement and engagement is that, as a leader, you will consciously make the choice to involve others, while your associates will make the choice – both consciously and subconsciously -- to engage on their own. The effectiveness of your involvement efforts will determine your ability to "pull" associates in, and its success is dependent on creating a compelling level of buy-in. Both the buy-in and the "pull" will materialize as an outcome of the way you manage, communicate, and stimulate the flow of change.

The motive to engage will take shape in the early stages of your improvement initiative through the methods you choose to reduce resistance. Anticipating the causes of reluctance to participate, simplifying the technical message in order to neutralize complexity, and positioning all communications from the perspective of your associates will allay their concerns and enlist engagement. You can close the deal by carefully defining the benefits of the effort in terms of what's in it for them, as well as what's in it for everyone. Keeping associates connected to the motives behind and benefits of the changes while integrating management support across all activities, positions you to advance the engagement efforts.

Standardize to Simplify and Sustain

Now, let's revisit the Lean tools, specifically standard work and Jidoka. As we learned in chapter seven, standard work insists that everything you do to add value for your customer can be broken into three simple elements: major step (what to do), key points (what to watch out for) and reason why (why it's being done). All processes have unique components, but also contain similarities, and the use of standard work creates an opportunity to leverage those common components through consistency of approach. Standardization has three by-products which will substantially improve the organization's ability to sustain its improvements. First, the process commonalities make it easier for associates to learn across all of the processes, simplifying training and reducing the amount necessary. This enables associates to move more fluidly between processes while remaining effective.

125

Second, standardization reduces the amount of communication needed when changes are introduced to the processes, making it easier to keep them current. Finally, leveraging these commonalities can facilitate a more rapid rate of improvement due to the synergies created across the business.

Jidoka, on the other hand, attempts to develop a habitual set of responses to business issues through the use of specially designed corrective tools. These are Lean based tools capable of counteracting or preventing performance abnormalities that can generate waste, deteriorate productivity, and threaten customer relationships through performance losses. Early stage Jidoka techniques are highly reactive, often focusing on containment steps within their responses, but they should evolve into preventive measures if given the correct attention.

Standardization is a difficult tool to stay abreast of, if for no other reason than it can run counter to our natural desire to create, refine, evolve, and improve the business, actions which always generate isolated solutions that can break down the standards. Keeping improvements focused and connected across the organization so that they don't dilute its benefits takes tremendous effort, but the lesson is, that standardization is a simplifier, making your processes easier to live with and sustain so long as the drive for improvement doesn't allow complexity to creep back in.

Auditing for Sustainment

Perhaps attributable to its more traditional uses, auditing is one of the most underused and/or misused practices in business. Surely there are financial and human resource professionals out there saying, "He must be out of his mind!" But if you've stuck with me this long, you already know there is a twist. For our purposes, let's simply define auditing as "verifying the performance of a process against its requirements." Although that definition makes it sound more like a glorified version of inspection, auditing differentiates itself by its more random frequency and a focus on more than just the results of a process. As well, when auditing is extended into the process functions, the results can be assured to more consistently meet the standard applied to them.

We will often relentlessly audit financial results and compliance history, always with a focus on how accurately these processes adhere to the rules. The problem in these examples is that the results-based review timing combines with the compliance mandate to create a negative connotation (the "gotcha" effect) for those being audited. Further, the real world impact is that the compliance activities associated with an audit are often crammed into the last few days prior to the event taking place, elevating the stress and fear already present in the team due to nature of the audit process, its feedback, and in many cases, the consequences. For our uses, turning auditing from enforcement to reinforcement transforms it into a very positive motivator in the pull behavior tool kit that changes a culture and aids it in sustaining improvement.

Positive Intervention – For several years, I've had the pleasure of working at an oil refinery in the Canadian Oil Sands Region. No organization takes safety as seriously as they do, and for good reason – the risk of failure can be catastrophic. These folks have put together millions of work hours without a serious injury, and two of their cornerstone techniques involve behaviors and auditing.

Behaviorally, they utilize the concept of intervention. When someone is observed working in an unsafe way, it is the observer's response-ability to step in and intervene to address the problem. In order to surround that effort with a positive theme, they expend a considerable amount of energy coaching everyone at every level of the organization as well as the contractors they hire that it's okay to intervene, establishing an expectation that even negative interventions can be delivered in as positive a fashion as possible. There is yet a further expectation on the part of the recipient that a respectfully delivered intervention is received respectfully. The entire site supports the practice with a zero tolerance policy for any lack of courtesy surrounding safety issues.

Although negative interventions must occur immediately as discovered, each member of the leadership team is required to counter every negative intervention with four positive ones. In a positive intervention, a worker is proactively thanked for working safely, performing a safe act, or for assisting someone else to work safely, with some kind of reward, ranging from a sticker to a keychain or even a pocket knife. The intent of establishing the ratio is to drive a positive overall tone to the program, increasing its acceptance.

The intervention approach is supported by an intense auditing methodology. All members of management are tasked with daily audit responsibilities for safe work practices, as well as personal protective equipment. They

Lowell J. Puls

report their findings, maintain logbooks, and generate daily grade cards for the leading indicators of risk-tolerant behaviors, following them up the next day with site-wide awareness initiatives designed to address the largest problems, getting the attention of everyone involved. The auditing data is used to feed a series of predictive metrics, which are aggressively acted upon.

Establishing a quota for positive interventions is the magic bullet that moves their safety approach from a reactionary one with a negative tone, to a proactive and positive one. It's a requirement that has the side benefit of building engagement by "pulling" the behaviors of everyone involved.

Reinforcing Pull Behaviors

Remember the concept of pull behaviors introduced in chapter six? There, we were trying to identify and solicit those behaviors that will directly support (and therefore sustain) an improvement. These are required to close gaps in three areas: process execution, abnormality response, and team based problem solving. A thorough job of standard work should assure process execution by virtue of its content, and Jidoka techniques will set in place fundamental practices for providing competent response to abnormalities. Sustaining team-based approaches for initiatives that range from launch through implementation, helps to establish a forum for peer to peer coaching and layered performance reviews, both triggered by auditing activities and metric checks.

Establishing audit points for key performance indicators at the beginning of your change initiative can help close gaps in process execution. Examples of this include audit points for incoming service calls, materials receipts, receivables, customers in line, and elapsed time through a process step. Periodic monitoring of process inputs provides a clearer indication of the need for preventive or corrective action, whereas auditing abnormality responses can help to confirm that process problems are correctly resolved within the necessary response-cycle. Some examples might include dropped sales calls, generation of nonconforming product, machine stoppages, late order fulfillment, and tardy payables.

Team-based problem solving opportunities can be identified during project reviews, sourced from customer-submitted issues, or discovered during the course of process execution and abnormality response. The completion timing and quality of problem resolutions also offer auditing options which can yield peer-to-peer coaching opportunities and include performance feedback.

Auditing alone doesn't create pull behavior, but will strengthen and accelerate it substantially when performed with the proper timing and intent. Turning the audit function from a practice that exposes a failure to perform within the rules, into a proactive approach that emphasizes positive feedback and timely coaching will make it both constructive and instructive. Process failure becomes replaced by preventive actions, while individual failure is avoided through communication and training. Emphasis on that last element of standard work, the reason why, becomes the glue that encourages associates to stick to the process rules that help to sustain them. The list of things done poorly metamorphoses itself into an inventory of finely tuned improvements, and it's all an outcome of simply taking a different angle on how you present it and look at it. The farther along in the process you extend auditing, the more successful the preventive efforts will be in lowering the number and the severity of the issues experienced.

Expanding the audit base

Broadening associate involvement in auditing throughout the organization has two benefits. First, more auditing is performed across the spectrum of activities, adding capacity for proactive process reviews. Second, with more participants, each audit can become smaller in scope, making them easier to perform effectively. A common deployment option is "layered auditing", a technique which refers to audits that are performed by all levels of the organization (from group leaders to senior managers). The positive aspect this approach is that it engages all levels of the organization in obtaining a multi-functional perspective of performance. Its greatest limitation is that the breadth of auditing will be restricted by the amount of resources committed to the program, and the higher levels of the organization may participate too infrequently to have the desired impact. Layered auditing only reaches maximum effectiveness if every organizational level participates to the extent necessary to achieve the program objectives.

The first requirement of broadened audit involvement is for each employee to perform some type of audit function. It should be limited to something within their capabilities – an example might be as simple as area 5's (workplace organization) compliance to plan; but all associates must play a part. It's the participation that helps create the momentum for cultural change, bringing with it frequent reinforcement for understanding the organizations goals.

An added benefit comes from the expanded capacity provided to the audit program, which allows it to broaden its role beyond compliance into both preventive and sustaining activities, thus accelerating the impact of the improvement programs.

Team based auditing can be used to maximize management effectiveness by narrowing the scope for critical aspects of process support. A number of approaches for audit team structure are possible, but centering them either on natural functions or even the specific strategic initiatives are excellent starting points.

By taking the extra care to proactively audit your processes and deliver the findings constructively, while providing support systems for the peer-to-peer positive interventions and coaching, the road to success will be far easier to travel.

Lean Assessment

A sizeable part of my career has been spent developing and launching Lean Assessment programs. One of those companies had established its primary objective for Lean Assessment as a tool for measuring and enforcing compliance to the corporate Lean implementation schedule. It had taken that turn because of the reticence a few of its key executives had shown in embracing the corporate Lean initiative as fully as they should. The assessment process was initiated at the CEO level to accelerate the adoption of Lean practices across the corporation. A requirement to increase an organizations minimum "Lean Assessment" scores was imposed first in the manufacturing facilities and then in the administrative offices. It worked well enough to improve the behavior of some of the executives in question, but didn't help Lean take root on a grander scale. Although the assessment process had done a great job of evaluating Lean tools usage, it failed to address the cultural behaviors that needed to change, leaving a lot of room for transformational improvement.

Similar to some audits, the findings from an assessment process simply reflect a snapshot of the processes current state of effectiveness. As I looked through the second draft of my fifth iteration of Lean Assessment (reviewing the point scores and criteria to see if they were properly aligned), I was still looking for a way to make the assessment feedback more useful. It was then I realized that an assessment shouldn't simply provide a score, but rather a set of precise recommendations for the subject's improvement. If the assessment identifies key areas needing improvement, then the evaluation criteria itself can serve as a path to improvement or a roadmap. With that the audit results were turned into a step-by-step guidebook to explain how to accelerate the Lean transformation. That change in its "angle of approach" moved my Lean assessment beyond benchmarking to become a template for evaluation and action planning.

Positive Power

As stated before, the power and value of auditing grows significantly when it's spun toward positive intervention. It becomes a support mechanism for behavioral engagement with some very effective improvement tools as its outputs: feedback, involvement, and forward planning. Like the Lean tools, auditing must be embraced by the organization and deployed broadly through "layering." This ensures that managers are always in a contributing auditor role rather than just a leadership one, facilitating improvements to their processes instead of directing outcomes. If the output of an audit is used to develop a road map to reach the next level, it will be more constructively received than if simply presented in the form of a punch list of deficiencies.

Finally, the value of engagement should never be discounted. Not only does broad audit engagement help accelerate the positive contributions, but it can also develop technical and leadership skills while deploying a more uniform set of expectations for all.

Summary

Sustaining your gains is essential to the successful evolution of any improvement initiative, but it's the most difficult aspect of every one of them. Unless they are properly dispatched, old behaviors will creep back to replace the new ones. Facilitating sustainment is a function of maintaining a pace of positive momentum, using creative methods to involve and engage, em-

Lowell J. Puls

phasizing standardization, and building audit systems that engage broadly in support of the changes by encouraging the desired behaviors.

You can further enhance the effectiveness of audit processes by acknowledging positive contributions. Rewards, recognition, and celebrations are powerful reinforcements that provide sufficient motivation to build team-based action and accelerate progress.

Chapter 10:
Consistent-ize
Business Systems that Secure Long-Term Performance

1) Defining Core Processes
2) Confirming Process Performance
3) Assign Process Ownership
4) Aligning Metrics

"Welcome, folks!" Geoff was batting leadoff for the staff as he addressed the team. "I have the pleasure of getting our first business transformation celebration underway, and am going to ask my teammates on our staff to set the tone by reminding you of our accomplishments during the past year."

Dick rose and spoke, "Not only have we saved our account standing at Valhalla Corporation, but we've actually grown our business with them by nearly 15 percent. We've redesigned four major platforms and successfully launched two of those with enthusiastic customer acceptance." Murmurs of approval were heard around the room. "In addition, we have reduced our engineering order-processing time from seven days to just two."

Geoff stepped back up. "Our safety incidents have dropped more than 75 percent and our goal of zero is in sight. We have also driven delivery up 15 points to 91 percent."

"And we've done it with one-third less inventory," Janet added.

Mary moved to a podium. "Fifteen percent of our associates have completed a developmental course, and eight percent have improved their performance levels, based on their end-of-year assessments."

Lowell J. Puls

Karen added, "Our profitability has improved nearly four percentage points in what many agree was a down market!"

I returned to the podium. "I'm proud of all of you for all you have accomplished, and the truth is, we've only started on our overall transformation. I commend each of you for your focus, your determination, and your loyalty. Everyone in this room, and across the entire company, has made a positive contribution to this amazing turnaround. Well done!"

I led the group in giving itself a round of applause.

"Naturally, we can't relax or rest on our laurels and assume the job is done. For in fact, improvement is a never-ending challenge, but it is self-rewarding. Knowing this, my goal is to guide you to improve on what you've already begun to do well. This morning, I'd like to review our primary process initiatives for the coming year. By doing this as a group, everyone will understand how critical these are for our future success."

I shifted gears for a moment and hefted a sheath of handout pages that would be distributed to them.

"To develop plans and objectives for next few years, the staff and I assembled a larger group of the management team to ask customers for their input, and to look within the corporation for benchmark opportunities. What we came up with was a list of core processes that we'll use to accelerate our transformation across the business by improving each segment of our company. The individual members of the staff will now present an overview of the core process that he or she will champion. I'll lead a process simply called leadership. The goal of a leadership process is to improve the way we establish the long-range and short-range planning for the business, making sure all of the appropriate input is considered. We'll call ours PDMR, which stands for Plan-Deploy-Measure-Review, and I'll have more information for you on the process in the next couple of months.

I then waved my hand to indicate that each person around the table should explain their individual roles in the overall improvement effort.

"I'll be the champion for our strategy deployment process along with how we develop and implement it through our annual action plans," said Karen.

"My process work will focus on how we listen to our customers and use their input to improve," explained Dick.

KISS: Keep It Simple and Sustainable

Mary said, "My team will first work on how we evaluate our performance, then determine how to use the feedback to drive personal development."

"I'll be directing our product development processes," said John. "We've already done a lot of brainstorming, and it's definitely going to be an exciting year."

"Performance of our manufacturing and service processes will be my area," Geoff added.

"Finally, my team will take us to world class levels of market and customer response," Janet promised. "If there's one thing we've learned this past year it's that if the customers aren't satisfied, nothing else really amounts to a hill of beans. So, my team and I intend to become even more fanatical in our efforts to serve customers through improved delivery response time, reduced costs, and higher quality."

I nodded approvingly. "You have all done an excellent job. Each of these initiatives will remain focused on a single process and a few others that feed them. We'll incorporate changes as they dovetail into the overall improvement initiative."

A hand went up. "How are we going to do all of this around our normal work?" One of the young engineers asked.

"Fair question," I said. "We believe that by developing new processes which are leaner than the old ones, we'll take a lot of work out of the current ones. Eventually, workloads may actually be reduced. Any other questions?"

Nobody offered a response. "Since no one wants to stand between you and the party, let's get it going." With that, I excused them and we moved on to celebrate a transformational year.

* * *

Under normal circumstances, an organization's natural ability to initiate and absorb change won't be rapid enough for it to adapt and avoid collateral damage when its own markets experience upheaval. That's the time a change initiative becomes necessary to reduce your vulnerabilities. As explained in the concept of cycles of response, the failure to accelerate your rate of change to match the transformation taking place in your markets can be catastrophic. A lesson painfully learned during the financial crisis of 2008 is that size won't help. Some very large corporations folded their tents due to over-exposures that were exploited by unprecedented market conditions.

Many businesses are complex. They have products, processes, markets, customers, supply chains, computer systems, organizational structures, development plans, facilities, and even communities that can all contribute to turn something that may have begun simply, into an entity that is much more convoluted. Well intended people in one area of the business will implement corrective actions that impact other areas of the business. The impacted groups then apply ensuing corrective actions, which may alter the effectiveness of a different segment of the organization. The cause-effect chase snowballs until a systemic problem develops. Stop-gap fixes can't remedy the progressive disruption so, a total realignment is required.

It might seem as if focusing on consistency (consistentize) is a strange way to close out a book on simplicity, but taken in order, identifying your key processes and modifying them to support the changes imposed by your initiatives is the perfect way to insure sustainability. Consistentize was saved until nearly the end, not because you won't introduce stabilizing behaviors throughout the course of your process improvement initiatives, but as you begin to produce results you'll want to keep them coming. In this chapter, we'll apply all of the previously discussed sustainment methods to the way you structure and execute your processes.

As implied, the real sustainment methodology has been assembled step-by-step throughout the book. The best results spawn from well-constructed and tightly managed processes, and it's the integrity of their design that allows sustainment to build. We've talked about how to develop sustaining behaviors that are supported with visual process triggers, requesting action in ways that trained associates can understand. The consistency of response is provided by the training and reinforced by multi-level metrics which are audited by the entire organization. This holistic approach is completed by an expectation for results that is deployed organization-wide to build performance discipline – that's what it takes to sustain.

Driving change while expecting consistent results might seem like an oxymoron; the two forces seem to be in opposition. The reality is that you will typically invoke a change initiative because you want to create consistently improving results, and although there might be an occasion where you will accept a short term performance loss to gain a potentially significant

KISS: Keep It Simple and Sustainable

improvement, it usually isn't necessary. If you're driving change with an expectation of stable results, then you're most likely driving the wrong kind of change.

There are risks inherent to any kind of change initiative, because they disrupt embedded, comfortable behaviors - people's habits; how they think, act, and respond on a daily basis. Obtaining consistently improving results is a matter of finding out what must be changed to drive improvement and then focusing on *only* those things while at the same time protecting and leveraging any good embedded behaviors. Alfred North Whitehead once said, "The art of progress is to preserve order amid change and to preserve change amid order." His statement supports the idea that a leader must leverage the good and focus the change effort on that which needs improvement; anything more is unnecessarily disruptive and will certainly cause *waste!*

Early in my career, I believed strongly in "change for the sake of change." My reasoning was that it could be useful as a force to break people out of their daily habits or comfort zones. I often used the example of placing an obstacle in the employees' entrance that would force them to choose a different path to their work stations, getting them to think consciously about something they otherwise would have done reflexively. Throughout the years my temperament has altered, helping me to realize that change which isn't supported by tangible needs can damage the credibility of the change effort. The best approach is to introduce change where it will do the most good without wasting resources or unnecessarily creating angst (resistance). If you fail to do so, the behaviors you want to disrupt will become the ones that function as resistance to the changes.

No matter how strong your sustainment methods are, they will be ruined by inconsistent management or systems that don't perform to expectations. Getting consistent improvement from a change effort starts with a set of critical core processes that will serve as your management "constant." Let's define them for now as the processes that exist in nearly every business; taking customer orders, fulfilling their needs, managing cash, and developing employees. Said another way, these processes control the pace of day-to-day performance by the way they "Consistentize".

Define Your Core Processes

How are core processes identified? Often times they become defined intuitively based upon the business type. Some processes are driven by industry standards, and some are even mandated through regulatory compliance. The folks that developed the application process for the Malcolm Baldrige National Quality Award, sought a more standardized methodology that is focused around the needs of the business and its stakeholders.

The Baldrige process seeks to evaluate a business from the following seven perspectives: leadership, strategy, customers, business systems, human resources, product / service delivery, and results. It asks specifically how you go about developing and using the processes under each category, and it would seem logical that at least one key process for each should exist. That's actually a great place to start, except that it's possible to have more than one critical process under each of the seven categories, and every process doesn't have to be a key to the business. Key processes should be designated as such because the way you manage them can have a make-or-break impact to a business.

Building a process around the leadership activities that you perform each day might seem odd at first, but it's the most appropriate way to build consistency into your efforts. When a leadership team consistently manages the way it sets priorities and the expectations regarding them, it conveys a message of discipline, integrity, and dependability to the entire organization. An example process that we have used in seeking sustainable leadership results is PDMR: Plan, Deploy, Measure, and Review. The planning step should occur at three levels: strategic, annual, and organizational. Linking the deliverables among all three will help to connect your strategies to your tactics while shaping the organization to deliver the strategy's needs. The results phase closes an improvement and feedback loop for the entire process and enables synergetic adjustments, as PDMR is deployed through the use of strategic planning sessions and progress reviews as well as in the organizations development processes.

To accomplish this, a process similar to the one used in my proprietary Business 5s Process (B5s) will move you beyond planning into execution. In an indirect fashion, much of the simplicity of B5s has already been described within the context of this book. The first "S," Strategy, derives its approach

KISS: Keep It Simple and Sustainable

from the widely accepted "Strategy Deployment" process, which engages the organization in developing or refining its market and business strategy. The strategy is then divided into initiatives and metrics assigned to them, before being deployed cross-functionally. Generally it takes two processes to fully cover your strategic needs: a) strategy development which covers multiple years and, b) tactical initiatives deployment (TID), which deploys and monitors strategy attainment objectives for the current year.

The second critical process seeks to ensure that all customer needs are fulfilled, not with an ordinary sales process, but rather with one that will take a broader view of the satisfaction of customer needs. For our purposes, the requirements/ideas/feedback (RIF) process asks how the design of products and/or services actually considers and then fulfills customer needs. The process structure first necessitates a method to identify customer needs, research the appropriate market benchmarks and establish internal specifications. Next, comes an ideation phase where other stakeholder needs are added in. Finally, all sources of customer feedback should be constantly recycled to ensure that your metrics for their satisfaction remain relevant, and its best if you triangulate between the internally defined list of market needs, the list of customer defined requirements, and with the mutual set of objectives generated for the business.

Once you have formulated your change initiatives, they should pass through three filters. The first and most heavily weighted filter is that of customer requirements; anything that doesn't address customer requirements should be considered expendable. There are reasons for working on something that doesn't satisfy customer needs in the short term, such as developing a product ahead of the market, but clear justification is required for them to receive priority for resources before you commit them. The specific needs of the business become the second filter. Customer programs must be appropriate for the needs of the business, and where they can't be, special scrutiny must be applied. The third filter is market and regulatory requirements. It's important to ensure that competitiveness is defined by the market rather than a single customer, and any regulatory considerations must be clearly understood and resolved.

Business Systems

In the context of the Baldrige criteria, the processes associated with the seven categories comprise a complete business system. After establishing processes for leadership, strategy, and customer needs, the next process requirement is reserved for the management of critical information, sometimes referred to as information technology. The sub-categories might include order management, planning and inventory controls, time management, organizational communications, and accounting systems. The complete list is indigenous to the specifics of any particular business, but they should combine as a whole to enable you to manage market response and support your organizational Response-Ability.

Keeping associates performing at their best goes far beyond simply assembling a set of metrics with a review process to manage them. Improving the skills of your human resources necessitates a comprehensive organizational development process that cross-checks against the business needs and addresses multiple aspects of performance. The Business 5s process for associate development is called ADAPT: Associate Development and Assessment Process + Training. ADAPT ties objective and metrics performance to leadership, technical, and interpersonal competencies with disciplined performance review cycles to generate developmental actions. The process branches into separate training and talent development processes at the next level down.

Of all the business process categories, the delivery of product and services covers the most ground and with the most variable scope depending on the type of business. Some process examples might include; product/service development, launch management, new accounts closure, perishable goods reordering, sales and operations planning (also referred to as S&OP and which I was once convinced stood for; Surprise! and Operations Panics!), retail inventory verification; Lean Enterprise engagement; and many more. The list of possibilities is only limited by the diversity of the businesses utilizing the approach. Your final process list should focus on the specific nature of your business then align with the timing and actions of the rest of your processes.

Managing a market response model is something very few organizations seem to think about, but it can make a vast difference in the success of any business. Initially, it requires blending customer requirements with market benchmarks to find the competitive response target for a given market. Once you have defined your customer response needs, cycles of response can be employed across all aspects of the business to make sure that the individual processing and operating cycles are capable of matching those needs. From that point, emplacing a management process that utilizes timed review cycles which align with market needs helps to sustain a competitive edge.

Finally, having dedicated so much of this book to the topic of metrics, the last category of results refers to the methods used to monitor and prioritize activities in order to leverage the improvement trends generated through the other business processes. For my purposes, managing results is simply an outgrowth of the way you review and deploy metric related actions. First, remember to use simplified and clearly defined versions of the headline $SQDC^2+T$ metrics at the organizational level: S (safety) or W (wellness), Q (quality), D (delivery), C^2 (cost), + T (time). Second, make certain your reviews are routinely scheduled and that all metrics and sublevel metrics are assigned to an owner or champion. Finally, every metric review should specifically include an improvement loop that utilizes process feedback to confirm problem resolution, as well as refine and improve the process that feeds that metric.

As you set the tone for process improvement, limit your specific instruction *only* to a mandate for simplicity of approach and solutions. You'll have to constantly work to keep the team focused on alignment of the processes to your objectives and the operating disciplines that build sustainment.

Confirm Process Performance

Achieving consistent results begins with thorough approach and deployment. To establish your approach, you should first review the results history and verify it reflects what's required for both the business and for your improvement initiative. Examine carefully the results' definition as well as the quality of the reported data. If adjustments to either one are needed, they should be made and incorporated into your approach. Take great care to retain and leverage all best practices as well as great performances – they will help to energize engagement through the momentum of excellence. Lastly, confirm the existence of both measureable feedback and actionable criteria to use in closing the improvement loop.

With the approach to getting results clarified, it's time to do a deployment check. Broad deployment and participation in metrics improvement is essential to ensure organizational alignment and objective achievement. All segments of the organization should be equally enmeshed in attaining the objective set, because just as a watch won't work if all of the gears don't fit precisely, neither will your team.

Extend the positive momentum by publicizing your significant team and individual contributions in a way that galvanizes the effort. Any progress reporting behaviors that might be confusing should cease. As well, all deviations from process, reporting, and performance standards should be limited by defining them in terms of "when to" and "when not to." Finally, when rewarding behaviors, make sure you separate innovation from achievement and attainment. Though all are valuable, they truly have separate effects on the organization, draw on resources separately, and generate rewards on different timelines.

Define Process Ownership

Previously we covered the need to assign owners or champions to each of the seven (or more) key business processes. This was referring to the need for both functional process owners as well as leadership champions. Many processes will already have a natural functional owner: payables falls under accounting, sales aligns to sales and marketing, production attainment belongs to the Operations group, and so on. Other processes can have mutual functional ownership and may default to the function which has trumping regulatory or other specific requirements. Functional process owners are assigned responsibility to refine and monitor process performance with the assistance of a key process team, and to adjust the key processes definition and structure to maximize its effectiveness.

The leadership champions play a somewhat different role. They are required to perform high level performance reviews as well as process audits to confirm functionality and effectiveness. They will also serve as facilitators to resolve process issues that require external involvement from senior leadership or even customers. The leadership champion reviews and ap-

KISS: Keep It Simple and Sustainable

proves process improvements, while providing the team with business level feedback.

The cross functional team for each key process champion is made up of the process leaders /owners for each of the applicable sub-processes. This group should use business process improvement (BPI) techniques to initially map and document the process flow and the relationships of the sub-processes, confirming that they satisfy all prescriptive requirements such as regulatory, corporate, or customer compliance. The sub-process leaders should also conduct bi-annual or annual reviews of process effectiveness utilizing performance data, interviews of customers and key users, audit results, and the use of feedback to drive improvements.

There are two vital benefits of the key process management approach. First, cross-functional team involvement extends organizational alignment around the transformational objectives by virtue of the engagement and broader deployment. Second, the assignment of process ownership and the reviews incumbent to it, serve to close the improvement loop for each business process It's this review-and-improve loop that allows the processes to evolve and remain effective.

Align Actions and Metrics

The need for a business-level result is usually what defines a key process and many of its elements, and this book has thoroughly stressed the need to establish and maintain alignment between activities, results, and time because of their inter-dependencies. That said, it should come as no surprise that each key process and sub-process will require at least one metric to ensure that its effectiveness is represented by a result. Process-specific metrics help steer the actions of the improvement teams, and ensure their effectiveness and success. While most of these metrics will evolve from the processes themselves, others may come through triangulation along with some additional intervention to derive a useful measurement. The teams and their leadership champions will perform this task most effectively.

Lowell J. Puls

Summary

So, I've covered a lot of ground under the theme of "consistentize." Whether or not you use a Baldrige style approach or a different one, identifying the key processes that drive your business is an important step to solidify the improvements you are realizing. Follow it up by integrating the effects of your change initiative into their process structures, engage your associates in managing them, and install metrics to monitor their progress; and you will close the loop on consistent improvement.

I'm sure some of these concepts may sound more complex than they really are, but it's really their simplest forms that are most powerful in application. Advocate a focus on core processes, measuring only what you need to, engage everyone to understand how the metrics drive business performance, and use it all as a body of knowledge to manage the individual pieces at the right level to sustain the simplicity. Once these skills are mastered, your team will be more readily engaged and able to keep up. That's what makes it possible to preserve the order amid change and drive change amid order.

Chapter 11:
Connecting the Dots
Adding it all up for Extreme Business Excellence

Although at this point we've covered everything that was included in the original outline for the book, I'm sure that simplicity has appeared elusive at times, since it seems as if every situation tries to self-complicate. In this chapter, we will connect all of the individual concepts into a contiguous approach by reviewing and summarizing the lessons.

Assess and Engage!

Begin the engagement process with a careful situation assessment, involving your team in both the assessment activity as well as the establishment of early improvement objectives. Whether you're a newcomer to the business or a seasoned veteran, their collective understanding of its needs are superior to your own "one-eyed" perspective, and they must participate in the identification of potential solutions. Their participation is a key to facilitating "buy-in" and ensuring an effective implementation.

The term Response-Ability was created to emphasize the importance of being responsive to the greater needs of the business. As the senior leader, you have both the duty as well as the capability to align the entire organization in pursuit the needs of the business. The initial step in that effort is to get the leadership team to admit to the real condition of the business – a "confession of reality." It's the only way to get associates working on the most

145

Lowell J. Puls

urgent problems and also serves as an initial motivator – their contribution to a successful outcome.

Broadening involvement to everyone in the business expands the work "enterprise-wide," and enables comprehensive process solutions that can better resolve the business issues while attending to your customers or markets with maximum effectiveness. First, you'll need to understand and address the many things that distract from their commitment – individual influences that source from outside and inside the organization. Your path to success builds from consistency of leadership, as it relates to setting expectations, identifying approaches, and maintaining engagement; in all other words... organizational focus. Engaging the entire organization in the development of a vision keeps it from being "yours" and is a critical aspect of its credibility. It takes more than that to get the vision "rooted" to reality however.

Prepare!

"Rooting" the vision to reality makes it appear real, more attainable, and far less the "hallucination" that it might seem otherwise. That appearance of reality comes in part not only from the preparatory information used to frame it, but as well from a carefully planned approach to its achievement that combines a realistic pace of activities to reach the improvement milestones that encompass all of the business needs.

The more broadly you deploy your change initiative, the more likely its chances of success. Assigning initiative related activities engages associate participation, and builds interest by virtue of the expanded involvement. Keep them informed of changes that are driven by external forces, as well as your adjustments to them. Continually updating results helps maintain their interest throughout any "course corrections" your change initiative might require.

At its outset, you'll have to estimate and understand the impact of the changes on your organization. How significantly are you altering the core business? Will customers be impacted? How many people are affected internally and in what ways? Answering these questions will give you an idea of the "mass" you're attempting to move, and how much force (resources + engagement) will need to be applied.

Successful change initiatives aren't just launched, but rather they are the product of thorough preparation. Sizing up the potential impact of your

KISS: Keep It Simple and Sustainable

change initiative can be accomplished using directional estimates for the project workloads as well as the resources necessary to support the "bill paying" activities that must remain ongoing. In the same way a good football coach will prepare relentlessly for a game, you'll need to prepare for your change initiative. Just as the coach will do a thorough inventory of player skills, one that includes existing capabilities as well as other potential options they may offer individually, he must also try to understand where his personnel might not adequately match up with those of the competition, and make appropriate adjustments. Your team assessment should thoroughly cover resident skills and performance history, as well as identify any skill or performance gaps that need to be adjusted for through associate development or personnel changes.

Before rolling out your initiatives, you should inform the organization of the vision and its overall objectives. Their participation and commitment is built from a trust that derives its strength from broad involvement, accurate measurement and clear communication of progress, and celebrating their achievements against the goals, while allowing for active game plan adjustments to prioritize and mobilize resources.

Prioritize!

Establishing priorities during a change initiative is one of the most critical aspects to its success, and they are highly dependent upon the condition of the business at launch. A business that's performing poorly and at risk to lose key customers must focus very hard on the "Things that will Kill them" to avert catastrophe. At the same time, they must avoid getting caught up in a focus purely on survival or else the change initiative will be stillborn. Balancing between "Things that Kill you" and progress on your strategy is always difficult and requires careful, constant management of resources. You can free up internal resources by removing priority from the "Things that make you Happy," and redirecting them to the most critical initiatives. Next, take a serious look at those "Things that Eat You" to ensure they are receiving the right amount of attention.

Fully resourcing the "Things that Kill you" without over staffing them, ensuring the "Things that Eat you" are resourced appropriately for their severity, and moving the "Things that make You Happy" off of the list, will provide the best support possible to your change initiative. If that still isn't

147

sufficient, utilize outside resources anywhere that their help can break free those initiatives with the highest "system effect" or return on investment.

In addition to developing a focus on the right priorities, once you establish a game plan, it's important to stick to it and not abandon it at the first sign of trouble. While adjustments are fine, too much reprioritization of your initiatives can induce organizational schizophrenia. If you want your associates to support and follow you in achieving the business goals, make sure your own support for the plan remains stable and your course corrections clearly justified.

Thus far, you will have expended tremendous effort to get the change initiative off the ground, and your engagement factor should begin to swell. Keeping the entire team focused on objective achievement, while performing complementary team-based activities and responding to assist when things begin to fall behind is made easier by deploying a set of "Pull Behavior" tools.

"Pull Behaviors" will change your culture from within by using a standardized set of "triggers" to request a response based action (think of stopping at a stoplight) in the face of an abnormality. The three components of the system include the trigger or "request" for action, people who understand it's their responsibility and are trained to "provide" a response to the trigger, and a "Standard Work" set of defined responses that serve to "guide" the solution.

Pull behaviors create a team effort around the correction of performance abnormalities and it's this timely response that minimizes their damage, helping to drive the motives for "blame" out of the culture.

LEAN into your change initiative

The most powerful tool set you will find to implement and reap benefits from your change initiative comes in the form of Lean Enterprise. Regardless of the type of business; manufacturing, services, health care, etc., Lean has proven techniques to help understand your customer's needs and provide value added activities exactly at the rate they are required. Although Lean utilizes a series of older tools, it offers two critical differences that make it far more effective. First, it recognizes that the engagement of the entire team is required for success. Second, it uses tools that are inter-dependent, in other words, using them individually won't give you the impact required.

KISS: Keep It Simple and Sustainable

In the simplest of terms, Lean demands that you intimately understand your customers and market, and utilize its rate of demand or *takt*, to build systems that are able to respond cost effectively to that demand, or flexibly respond to changes in it. The tools will enable you to "flow" activities at *takt*, while pull behaviors will mobilize you to provide rapid response to changes. A well implemented Lean tool set can yield truly amazing results.

Monitor your progress!

A new change initiative deserves a revised set of business metrics to go with it. While that doesn't require you to change or eliminate all of the traditional metrics, the application of clear benchmarks and objectives will provide renewed focus. It's easy to overburden an organization by measuring everything under the sun, so keep your key metrics list simple, focused on business critical measurements, and maintain review cycles that are in line with your ability to impact them. As quickly as you can, try to evolve them away from looking solely at results and toward predictive measures that are more helpful in steering the business. The right set of metrics, combined with timely review and follow up, will serve to energize organizational engagement.

Each of the improvements that develop as a function of your change initiative will require stability and support for them to be sustained. Although we've spent a considerable amount of text offering engagement approaches for the activities around your change strategy, its thrust has been toward driving change and achieving goals. Most likely, all of us have worked in companies that were great at achieving an objective, but which weren't as good at sustaining the gains. Improvements will fall apart if we fail to understand the "maintenance manual" for the changes. A well designed process audit program is a useful tool to help guide the construction of that manual while at the same time using the increased associate involvement to broaden engagement. The system design activity should include participation from *all* members of the organization, and positioning the auditing topic/activities as far forward of the results as possible can yield the most positive impact.

Make Improvement Systemic!

The final element of sustainability is to develop a systemic approach in which all of the improvements can find their way into your business

Lowell J. Puls

processes. Doing this will necessitate identifying and separating your "core processes" from the larger group of all of your business processes. These core process lists will very likely follow common themes within different types of businesses, but should include at minimum processes that cover customer facing activities, development of offerings, delivery of offerings, associate development, management of information, and Leadership. A critical success factor in building sustainability is to assign a champion or owner to each core process and their critical sub-processes. They will assume responsibility for leading a team in monitoring results (metrics) and using them to actively evolve the processes effectiveness. This is the step that constitutes the "cement" of sustainability for your improvements.

That should summarize it all. While there is plenty of supporting detail to help you in each of the chapters, every unique business situation will serve as an individual test to many of these approaches. This is where numerous Lean implementations have served well, teaching me that while the details of the implementation will always change, they will equally be resolved by the details of the solutions. The simplicity of the high level approach is what should remain consistent in order to achieve the desired result.

Good luck sustaining your gains.

Addendum

"I'm not on board with the changes you're advocating and am instructing you to stop with them. Further, there will be no discussion of this with anyone at corporate headquarters. This plant is doing just fine without their help."

My boss, the plant manager was explicit, and my attempts to strengthen my point, only caused him to terminate the discussion abruptly. "Here is a copy of your self-appraisal. I'd like you to fill it out with your impression of your performance and return it to me by the end of the week."

I'd only been on his staff for just a few months and, unknown to me, had been imposed upon him by a senior human resources executive who wanted to stimulate some change from underneath – career avoidance advice for all of you newcomers. Our honeymoon hadn't been stellar, and we seemed to be at odds at every turn. He had built an accomplished profit machine, winning credit year after year for financial performance that always exceeded his business plans. Such consistency was the result of a number of different things, but a key element was his very firm control of everything within the facility's four walls. Any level of change was scrutinized heavily for its downside, and even the slightest chance of a negative impact would stop it until it could be reworked to be risk-free. The rate of change was agonizingly slow, and because consistency of results was what he had been so well-rewarded for, he was careful to protect it.

When I had completed the review and attempted to return it as instructed, he asked me to hang on to it, and continued to put it off until it was never used. He ultimately gained sufficient confidence in me to begin allowing me to experiment with some of the improvements that had been proposed. Still, because of the requirement to modify them until they were entirely risk-free, my progress was good but not great. The very next year, a

Lowell J. Puls

significant event occurred that completely altered the market dynamic. Our largest customer (and the market leader) entered into a supply agreement with our primary competitor, who had just been purchased by one of their larger suppliers. The agreement resourced many of the products we'd been providing them to our competitor, impacting more than 20% of our total business revenue. One deft move literally inverted our respective market positions overnight, triggered by that most powerful of market forces – price. Not only did we lose a huge segment of our market share, but we were also forced simultaneously to reduce our pricing on the balance of the products we sold to that customer.

The event was an apocalypse from the perspective of our business. Pressure to stabilize profitability in the face of such a radical market shift built instantly, and it opened the doors for a much more rapid rate of change than anyone could have imagined. Although it had taken this general manager 22 years to perfect his profit machine, it took only a few months and a sudden shift in the market to unravel it. In the end, he was left with no choice but to attempt some radical moves in response to the market, driving a level of change that he truly struggled with. Feeling frustrated, he retired in its midst.

The profit engine had fallen off the tracks, and suddenly there was a receptiveness to change borne of the urgency to survive. Among it all, a key element of the business model he had built – its responsiveness -- remained well-preserved, and it quickly helped us capitalize on opportunities to regain sales volume every time our competitor encountered performance problems, which happened frequently.

A key lesson for me was that consistency of performance is a huge asset to any business, and it originates with managerial behavior. The predictability of performance that it breeds -- consistency of expectations, regular achievement of metric targets, and careful evolution of operational processes - can also have the side-effect of stifling improvement. That general manager's fear wasn't rooted in the threat of change, but that changes of excessive magnitude or speed might compromise results. He had long ago come to understand a lesson that he would end up teaching me: you can only change as fast as the organization can follow. Later on, I added a counter balanced learning that the level of effort you expend in preparing people for change will return itself in an accelerated rate of adoption. He advocated using

KISS: Keep It Simple and Sustainable

proven techniques in a highly controlled environment and in a way that would move results forward without any chance of adverse effects. Although his techniques were brilliant, his own understanding of the behavioral impact was mostly one-sided and resulted in sub-optimization.

Ultimately, the business was able to re-stabilize profitability, albeit at a lower overall level. The market had made a permanent adjustment, commoditizing the product and forever losing its pricing premium. Our return to consistency was attributable in part to a strong commitment to Lean and also the same thing that had gotten us there in the first place; customer responsiveness and a commitment to delivering consistently excellent performance. It was a basic expectation within that business's culture that couldn't be derailed, even by a major market event.

Hopefully, the original promise to keep the concepts of re-energizing your business both concise and simple has been upheld. My friend's belief that business leaders are highly intelligent people whose tendency to overcomplicate problems seems accurate enough. Both their approaches and the solutions for many business issues too often exclude the human dynamic, causing confusion and failure. An essential lesson of leadership lies in your understanding that complexity makes it harder for associates to follow any approach, thereby reducing speed, impairing results, and potentially stalling improvement.

Addressing complex problems with simple solutions takes remarkable discipline. The fact is, the more effort you expend on simplification, the less effort will be required for the actual activities and contingencies. The better you get at it, the greater speed your initiatives will gain, benefiting from their superior ease of associate understanding and engagement.

Don't think for a minute that I've discounted or ignored the complexity that exists in many of the situations covered in this book. Having actually lived them, every scenario offered as an example was a real one that bore its own "how-to" lesson in simplification. The trick is to strip the problem to its basic needs and provide a solution for each level of process management. At the same time, you have to limit focus on the metrics at that process level and at the next level down (its first level pareto). The assigned teams will be tasked with timely response, avoiding the confusion created by trying to manage too many metrics.

Keeping the organization working within their range of influence

Lowell J. Puls

prevents the complexity that comes from accumulating multiple action levels and causes, onto the one specifically being worked. That's the essence of simplification -- not to ignore the complexity, but simply to isolate it at the most appropriate place for you to work it and not allow it to accumulate elsewhere.

All of the concepts I've introduced -- cycles of response, reconnoitering, triangulation and others - are intended to focus your attention on the highest level of specific inputs that can help identify problems, resolve them, and lead to positive actions and decisions. They serve as reality checkpoints at times when massive amounts of information tend to confuse and create disarray.

Change initiatives attract complexity, but don't allow it to stop you or intimidate you from getting started. The choices often made in order to preserve a current state of performance, while averting risk or performance loss, can frequently be detrimental to the long-term health of a business. Let's close with one more Machiavelli quote; "The wise man does at once what the fool does finally!" So, stop contemplating your change initiative and get started!

Lowell Puls

About the Author

Lowell Puls's manufacturing career has taken him from the shop floor, to leading more than $300M in global operation. His customer and employee focused style has contributed to the revenue growth of many companies, while at the same time delivering millions in profit improvement.

His undergraduate work in manufacturing technologies and education, Loyola MBA, and as an Executive Development graduate for a Fortune 500 company, Lowell has developed the skills to create and deliver transformational strategies - "changing the game" for great businesses as well as those needing help.

CPSIA information can be obtained at www.ICGtesting.com
Printed in the USA
LVOW08s2125111013

356579LV00001B/54/P